VIETNAM STUDIES

MOUNTED COMBAT IN VIETNAM

by
General Donn A. Starry

DEPARTMENT OF THE ARMY
WASHINGTON, D.C., 2002

Library of Congress Cataloging in Publication Data

Starry, Donn A 1931-
Mounted combat in Vietnam.

(Vietnam studies)
Includes index.
1. Vietnamese Conflict, 1961-1975—Campaigns. 2. United States. Army. Armored Force—History. 3. Vietnamese Conflict, 1961-1975—United States. I. Title. II. Series.
DS558.9.A75S73 959.704'342 78-12736

First Printed 1978—CMH Pub 90-17-1

For sale by the Superintendent of Documents, U.S. Government Printing Office
Internet: bookstore.gpo.gov Phone: toll free (866) 512-1800; DC area (202) 512-1800
Fax: (202) 512-2250 Mail: Stop SSOP, Washington, DC 20402-0001

Foreword

The United States Army met an unusually complex challenge in Southeast Asia. In conjunction with the other services, the Army fought in support of a national policy of assisting an emerging nation to develop governmental processes of its own choosing, free of outside coercion. In addition to the usual problems of waging armed conflict, the assignment in Southeast Asia required superimposing the immensely sophisticated task of a modern army upon an underdeveloped environment and adapting them to demands covering a wide spectrum. These involved helping to fulfill the basic needs of an agrarian population, dealing with the frustrations of antiguerrilla operations, and conducting conventional campaigns against well-trained and determined regular units.

It is still necessary for the Army to continue to prepare for other challenges that may lie ahead. While cognizant that history never repeats itself exactly and that no army ever profited from trying to meet a new challenge in terms of an old one, the Army nevertheless stands to benefit immensely from a study of its experience, its shortcomings no less than its achievements.

Aware that some years must elapse before the official histories will provide a detailed and objective analysis of the experience in Southeast Asia, we have sought a forum whereby some of the more salient aspects of that experience can be made available now. At the request of the Chief of Staff, a representative group of senior officers who served in important posts in Vietnam and who still carry a heavy burden of day-to-day responsibilities have prepared a series of monographs. These studies should be of great value in helping the Army develop future operational concepts while at the same time contributing to the historical record and providing the American public with an interim report on the performance of men and officers who have responded, as others have through our history, to exacting and trying demands.

The reader should be reminded that most of the writing was accomplished while the war in Vietnam was at its peak, and the monographs frequently refer to events of the past as if they were taking place in the present.

All monographs in the series are based primarily on official records, with additional material from published and unpublished secondary works, from debriefing reports and interviews with key participants, and from the personal experience of the author. To facilitate security clearance, annotation and detailed bibliography have been omitted from the published version; a fully documented account with bibliography is filed with the U.S. Army Center of Military History.

The story of mounted combat in Vietnam was written at Fort Knox between 1973 and 1976 by a task force under the direction of Major General Donn A. Starry, then commander of the Armor Center and commander of the Armor School. General Starry has been involved in the planning or direction of armored operations and development since he was graduated from the U.S. Military Academy in 1948 as a second lieutenant of cavalry. After serving in command and staff positions from platoon to battalion in armored units in Europe until 1953, he became a staff officer in the Eighth Army in Korea and then an instructor in combined arms and nuclear weapons employment at the U.S. Army Intelligence School. He later served as an armored battalion commander and staff officer in U.S. Army, Europe. In 1966 he assumed duties in the G–3 Section, U.S. Army, Vietnam, and was a member of the Mechanized and Armor Combat Operations, Vietnam, study group which evaluated armored operations in Vietnam. After serving in assignments with the Vice Chief of Staff of the Army and the Secretary of Defense, he returned to Vietnam to join the plans office of J–3, Headquarters, U.S. Military Assistance Command, Vietnam, and in 1969 assumed command of the 11th Armored Cavalry Regiment in Vietnam. In 1970 he returned to the United States and served successively as Deputy Director of the Operations Directorate and Director of Manpower and Forces. After two and one-half years as the commander of the Armor Center, he assumed command of V Corps, U.S. Army, Europe, in 1976. Promoted to full general, in July 1977 General Starry became commander of the U.S. Army Training and Doctrine Command, Fort Monroe, Virginia.

Washington, D.C.
15 September 1977

JAMES C. PENNINGTON
Brigadier General, USA
The Adjutant General

Preface

This monograph is an account of the operations of armored units of the United States Army in the Republic of Vietnam. The term *armored units* as used here is generic and includes tank and mechanized infantry battalions and companies, armored cavalry squadrons and troops, and air cavalry squadrons and troops—all forces whose primary *modus operandi* was to fight mounted.

Of necessity the story begins not with the arrival of the first U.S. armored units in Vietnam in 1965 but with armor in Vietnam during the years immediately after World War II. The generally unsuccessful experience of French armored forces in Southeast Asia from the end of World War II to 1954 convinced American military men that armored units could not be employed in Vietnam. It was widely believed that Vietnam's monsoon climate together with its jungle and rice paddies constituted an environment too hostile for mechanized equipment; it was further agreed that armored forces could not cope with an elusive enemy that operated from jungle ambush. Thus at the outset of American participation in the conflict and for some time thereafter, Army planners saw little or no need for armored units in the U.S. force structure in Vietnam. At the same time, however, extensive American aid that flowed into Vietnam after the French left the country was directed in part to developing an armored force for the newly created Army of the Republic of Vietnam.

It was not until 1967, however, when a study titled Mechanized and Armor Combat Operations, Vietnam, conducted by General Arthur L. West, Jr., was sent to the Chief of Staff and Secretary of the Army, that the potential of armored forces was fully described to the Army's top leaders. Despite the study's findings—that armored cavalry was probably the most cost-effective force on the Vietnam battlefield—there was little that could be done to alter significantly either the structure of forces already sent to Vietnam or those earmarked for deployment. By that time, constraints on the size of American forces in Vietnam had been imposed by Defense Secretary Robert S. McNamara and decisions on force deployment extending well into 1968 had already been made. The armored force of the Army of the Republic of Vietnam, meanwhile had been successful enough in fighting the elusive Viet Cong that U.S. armored units had been deployed in limited numbers, usually as part of their parent divisions.

PREFACE

From early March 1965 until the cease-fire in January 1973, U.S. armored units participated in virtually every large-scale offensive operation and worked closely with South Vietnamese Army and other free world forces. After eight years of fighting over land on which tanks were once thought to be incapable of moving, in weather that was supposed to prohibit armored operations, and dealing with an elusive enemy against whom armored units were thought to be at a considerable disadvantage, armored forces emerged as powerful, flexible, and essential battle forces. In large measure they contributed to the success of the free world forces, not only in close combat, but in pacification and security operations as well. When redeployment began in early 1969, armored units were not included in the first forces scheduled for redeployment, and indeed planners moved armored units down the scale time and again, holding off their redeployment until the very end.

In almost equal parts this study has drawn from official war records of armored units and personal interviews with men of those units. The monograph makes no attempt to document every armored unit in every battle. Nor does it list in detail the lessons that may be learned from the Vietnam conflict, although it does call attention to some. In so doing it sometimes isolates and focuses on the mounted combat aspects of operations that actually included many different American and other free world units. The reader should keep in mind that the author's intent is to tell the story of mounted units, and not to describe battles in their entirety.

Documenting this story of mounted combat in Vietnam was not a one-man job. Of the many people who helped, several deserve special thanks. Lieutenant Colonel George J. Dramis, Jr., director of the monograph task force, developed the first topical outline, assessed the historical significance of each bit of the wealth of information available, and ran the task force from day to day.

The members of the monograph task force, Vietnam veterans with firsthand experience, whose collective knowledge contributed to the continuity of the story were armor officers Major John G. Russell, Major Thomas P. Barrett, Captain Robert M. Engeset, Captain John L. Hagar, Captain Gerald A. McDonald, Captain Maurice B. Parrish, Captain Jeffrey A. Stark, Captain Calvin Teel, Jr., and Sergeant Major Christopher N. Trammell; infantry officers Captain Robert P. Antoninc and Captain John J. Strange; and Captain Dennis M. Jankowski of the Quartermaster Corps. The contributions of the Infantry School, particularly those of Lieutenant Colonel Wayne T. Boles, were invaluable. Without the good work of the administrative staff, Mrs. Pege R. Bailey, administrative assistant, Mrs. Jeanne Meyer, typist, and many temporary typists, this volume would not have been completed. Specialist 5 Elmer R. Adkins, Jr., and Specialist 4 Jack D. Travers, who repeatedly typed the final drafts, deserve special mention. Not to be forgotten are the many lieutenants on temporary assignment with the task force who painstakingly researched articles and located officers and enlisted men who had served in Vietnam.

This monograph is an accounting of the stewardship of the tank crewmen, mechanized infantrymen, armored cavalrymen, and air cavalrymen who had a hand in some of the more significant events in the Vietnam War. It was their devotion, professionalism, valor, and dedication that brought American arms to the conclusion decided upon and ordered by their commander in chief. In the end they left us a large legacy. The monograph is their story—it is dedicated to them.

Fort Monroe, Virginia
15 September 1977

DONN A. STARRY
General, U.S. Army

Contents

Chapter *Page*

I. INTRODUCTION 3
Influence of French Use of Armor 3
U.S. Armored Forces After 1945 6
Vietnam as a Field for Armor 9
The Enemy in Vietnam 10

II. ARMOR IN THE SOUTH VIETNAMESE ARMY 17
U.S. Advisers 19
M113's in the Mekong Delta 21
Reorganization and Retrenchment 24
Expansion of Armor in the South Vietnamese Army 30
Cuu Long 15 31
Time for Corrective Analysis 33
Improvements in Equipment 37
Enemy Reaction to Armored Vehicles 45
"Coup Troops" 48

III. GROWTH OF U.S. ARMORED FORCES IN VIETNAM 50
The Marines Land 52
Decision Making 54
Scouts Out 58
Ap Bau Bang 60
Deployments and Employments 63
Task Force Spur 65
Battles on the Minh Thanh Road 66
The Blackhorse Regiment 72
Mine and Countermine 79

IV. COMBINED ARMS OPERATIONS 84
The MACOV Study 84
Cedar Falls—Junction City 91
Mechanized Operations in the Mekong Delta 103
Route Security and Convoy Escort 106
Air Cavalry Operations 111
Other Free World Armor 112

V. THREE ENEMY OFFENSIVES 114
Enemy Buildup 114
First Offensive: Tet 1968 116
Bettle of Tan Son Nhut 118
Battle of Long Binh and Bien Hoa Area 123
Battles in Vinh Long Province 127

CONTENTS

Chapter | *Page*

Second Offensive ... 129
Third Offensive ... 131
Aftermath ... 136

VI. THE FIGHT FOR THE BORDERS ... 138
Changing Strategy ... 138
Armored Forces Along the Demilitarized Zone ... 139
The Sheridan ... 142
"Pile-on" ... 145
Rome Plows ... 147
Tank Versus Tank ... 149
Invading the Enemy's Sanctuaries ... 153
Securing the Borders ... 156
Pacification Efforts ... 161
Vietnamese Forces Take Over the War ... 164

VII. ACROSS THE BORDER: SANCTUARIES IN CAMBODIA AND LAOS ... 166
Early Operations Into Cambodia ... 167
The Main Attack Into Cambodia ... 168
South Vietnamese Army Attacks Continue ... 176
Secondary Attacks Across the Border ... 178
Cambodia in Perspective ... 179
Maintenance and Supply ... 181
Lam Son 719 ... 186
The South Vietnamese Army Attack ... 190
Air Cavalry and Tanks ... 193
The Withdrawal ... 194
Cuu Long 44-02 ... 197

VIII. THE ENEMY SPRING OFFENSIVE OF 1972 ... 199
Point and Counterpoint ... 200
The 20th Tank Regiment ... 203
Attack Across the Demilitarized Zone ... 205
The Rock of Dong Ha ... 206
The Enemy Attack in Military Region 2 ... 212
The Aftermath ... 217

IX. REFLECTIONS ... 220

Appendix

A. VIETNAM UNIT COMMANDERS ... 227
B. ARMOR RECIPIENTS OF THE MEDAL OF HONOR ... 238
GLOSSARY ... 239
INDEX ... 243

Chart

Page

U.S. Armored Organizations, 1965 52

Maps

No.

1. Geographic Regions, South Vietnam 8
2. Dry Season, South Vietnam (and tracked vehicles) 11
3. Wet Season, South Vietnam (and tracked vehicles) 13
4. Battle of Ap Bac I, 2 January 1963 26
5. Ap Bau Bang I, 11–12 November 1965 61
6. Battle of Minh Thanh Road, 9 July 1966 68
7. Battle of Suoi Cat, 2 December 1966 75
8. Expanding U.S. Operations, 1967 92
9. Ap Bau Bang II, 9–20 March 1967 98
10. Battle of Ap Bac II, 2 May 1967 104
11. Major Armor Battles, *Tet* 1968 117
12. Saigon Area, *Tet* 1968 120
13. Battle of the Crescent, 20 January 1970 158
14. Operations Into Cambodia, May–June 1970 170
15. LAM SON 719, 8 February–6 April 1971 188
16. Northern I Corps, March 1972 202
17. 20th Tank Regiment, 1–27 April 1972 207

Diagrams

1. Capstan and Anchor Recovery 39
2. Push-bar Extraction 39
3. Cable and Log Extraction 40
4. Block and Tackle 40
5. Tow Cables 41
6. Cloverleaf Search Technique 88

Illustrations

M24 (Chaffee) Used by French in Vietnam 4
Tanks Firing in Support of French Infantry at Dien Bien Phu ... 5
Viet Cong Soldier 15
South Vietnamese Reconnaissance Unit With 1939 French Armored Car 18
Armored Personnel Carrier M113 23
Balk Bridge Carried by M113 42
M41 in South Vietnamese Training Operation 44
M113 Damaged by Viet Cong Recoilless Rifle 46

CONTENTS

	Page
U.S. Marine Corps Flamethrower Tank in Action Near Da Nang	53
ACAV Moves Out to Escort Convoy	74
ACAV's Form Defense Perimeter	76
Engineer Minesweeping Team Clears Highway 13	80
Tank-Mounted Mine Roller Prepares to Clear Highway 19	83
Herringbone Formation	87
M113's and M48A3 Tanks Deploy Between Jungle and Rubber Plantation	94
Tanks and ACAV's Secure Supply Routes	106
OH–6A Observation Helicopter and Two AH–1G Cobras on Visual Reconnaissance	110
Troops of 1st Australian Armor Regiment With Centurion Tank	113
Tank and M113 During Enemy Attack on Bien Hoa, *Tet* 1968	125
M41 of South Vietnamese Army Advances on Enemy Positions in Saigon, May 1968	130
M113 With Protective Steel Planking in Action at Ben Cui Plantation	133
Preparing Night Defensive Positions Along Demilitarized Zone	141
Sheridan M551 and Crew	142
Pile-on Operation in I Corps	146
Rome Plows With Security Guard of M113's	148
Russian-made PT76 Tank Destroyed at Ben Het	152
Tank in Position To Provide Static Road Security	162
The 2d Squadron, 11th Armored Cavalry, Enters Snuol, Cambodia	173
M88 Heavy Recovery Vehicle Loads Damaged APC	185
Red Devil Road	190
ACAV's of South Vietnamese 1st Armored Brigade in Laos	191
Captured North Vietnamese T59 Tank	210
South Vietnamese M48 After Hit From Rocket	213
UH–1B Helicopter With TOW Missiles	214
ACAV Takes Position for Counterattack Near My Chanh	216
M48A3 Tank Explodes Bomb Set Up as Mine	222
Engineers Clear Trail of Mines in Cambodia	223
M578 Light Recovery Vehicle Works on Sheridan	225

All illustrations are from Department of Defense files.

MOUNTED COMBAT IN VIETNAM

CHAPTER I

Introduction

In the aftermath of World War II the French colonial administration returned to Indochina to resume control of French possessions. With it came units of the French Army, among them mechanized and armored elements. These units remained in Vietnam for more than ten years, until, in compliance with the Geneva Accords of 1954, the last French soldier left the country in April 1956. The experience of the French and their Vietnamese allies in those years had a strong influence on concepts developed in the South Vietnamese Army for the employment of armored forces. Their experience also influenced the thinking of American military commanders and staffs when the U.S. Army eventually set about deciding how many and what kinds of forces to send to Vietnam.

Influence of French Use of Armor

The U.S. Army in the early 1960's had very little information on the use of armor in Vietnam, and most of that came from French battle reports and the fact that the French Army had been supplied with World War II tanks, half-tracks, and scout cars. Although most of this equipment was American, made originally for the U.S. Army, there was little reliable information as to the amount, condition, and use of it in Indochina. In 1954, after four years of American aid, the French fleet of armored vehicles consisted of 452 tanks and tank destroyers and 1,985 scout cars, half-tracks, and amphibious vehicles, but this armor was scattered over an area of 228,627 square miles. By comparison, in June of 1969 the U.S. forces in Vietnam had some 600 tanks and 2,000 armored vehicles of other types deployed over an area less than one-third that size.

All the American equipment used by the French was produced before 1945. In general the armor was not fit for cross-country movement and because of its age was often inoperative. The logistical system, with supply delays of six to twelve months, further hampered operations by making maintenance difficult. Helicopters were not available in large numbers—there were ten in 1952, forty-two in 1954; all were unarmed and were used for resupply and medical evacuation. To the French command, impoverished in all resources, fighting with limited equipment over a large area, the

M24 (Chaffee) American Light Tank Used by French in Vietnam

employment of armored forces became a perpetual headache. Armored units were fragmented; many small remote posts had as few as two or three armored vehicles. Such widespread dispersion prevented the collection or retention of any armor reserves to support overworked infantry battalions. When French armored units took to the field, they were roadbound. Roads prescribed the axes of advance, and combat action was undertaken to defend a road and the ground for a hundred meters or so on either side. The enemy was free to roam the countryside. Since armored units were generally assigned to support dismounted infantry, their speed and ability to act independently, an important part of any armored unit's contribution to the battle team, were never used.

All these facts were duly reported by the French in their candid, comprehensive, and sometimes blunt after action reports. In the United States, because of restrictive military security regulations and a general lack of interest in the French operation in Indochina, there was no body of military knowledge of Vietnam. What was known had been drawn not from after action reports but from books written by civilians. Foremost among these was Bernard B. Fall's *Street Without Joy*, which greatly influenced the American military attitude toward armored operations in Vietnam. One series of battles in particular stood out from all the rest, epitomizing the French experience in American eyes. Entitled "End of a

TANKS FIRING IN SUPPORT OF FRENCH INFANTRY AT DIEN BIEN PHU

Task Force," Chapter 9 of Fall's widely read book traced a six-month period in the final struggles of a French mobile striking force, Groupement Mobile 100. The vivid and terrifying story of this group's final days seemed to many to describe the fate in store for any armored unit that tried to fight insurgents in the jungles.

Actually Groupement Mobile 100 was not an armored unit at all, but an infantry task force of 2,600 men, organized into four truck-mounted infantry battalions, reinforced with one artillery battalion and ten light tanks. Restricted to movement on roads, deploying to fight on foot, it was extremely vulnerable to ambush, and, indeed, a series of ambushes finally destroyed it. Because most readers did not take the time to understand the organization and actions of Groupement Mobile 100, its fate cast a pall over armored operations in Vietnam for almost twelve years. The story of this disaster became a major source for unfavorable references to French armored operations in Vietnam, and contributed much to the growing myth of the impossibility of conducting mounted combat in Vietnam.[1] In fact, the myth was so widely

[1] General William C. Westmoreland, the U.S. commander in Vietnam from 1964 to 1968, kept a copy of Fall's book on the table near his bed. He later said that the defeat of Groupement Mobile 100 was "always on my mind," particularly so during the early U.S. deployments.

accepted that it tended to overshadow French successes as well as some armored exploits of the Vietnamese Army, and it actually delayed development of Vietnamese armored forces. Unfortunately, U.S. commanders were to repeat many of the mistakes of the French when American armored units were employed.

U.S. Armored Forces After 1945

When World War II ended the United States Army had an armored force of sixteen divisions and many other smaller armored units. The bulk of this force—the divisions—included balanced, integrated, mobile fighting forces of armored artillery, armored infantry, and tanks. Concepts for employment of these combined arms forces recognized no limitations of geography or intensity of warfare. The combined arms idea stressed tailoring integrated mobile forces to the situation, taking stock of enemy, terrain, and mission. Mechanized cavalry units were formed to supplement these forces by conducting reconnaissance and providing security.

The employment of U.S. armored divisions exclusively in Europe and Africa during World War II caused many to conclude that only in those theaters was warfare with armor possible. U.S. military studies of armor in the war were based on accounts of combined arms warfare in Europe and North Africa. Most American experience with armored units in the Pacific, and later in Korea in the early 1950's, seemed to confirm the impression that armored units had but limited usefulness in jungles and mountains. The Army staff therefore concluded that while tanks for the support of dismounted infantry might be required, there was no possibility for independent large-scale combined arms action by armored forces such as those of the World War II armored divisions. It was against this background that the U.S. Army grappled with decisions on American troop deployments to Vietnam in the early 1960's. Combined with the misconceptions of the French armored experience, this reasoning caused most planners to conclude that Vietnam was no place for armor of any kind, especially tanks.

In the war in the Pacific there was slow, difficult fighting in island rain forests. No armored division moved toward Japan across the Pacific islands. Neither blitzkrieg tactics nor dashing armor leaders achieved literary fame in jungle fighting. It was an infantry war; armored units were employed, but what they learned was neither widely publicized nor often studied. To most military men the jungle was a dark, forbidding place, to be avoided by armored formations. Even after American military advisers began to replace the French in the country, the very name, Vietnam, conjured up an image of dense, tropical rain forests, rice paddies, and swamps.[2]

To American military conceptions of jungle fighting, the Korean War added some additional experience that weighed in the deliberations on troop deployments to Vietnam. Korea has a monsoon climate, a seasonal change in the prevailing wind direction, which is offshore in winter, onshore in summer. The deluge of summer rains in Korea is a reality that no one who served there can forget. The rains made mounted combat difficult, if not impossible. The extensive flooded rice paddies in the western Korean lowlands were added obstacles to the movement of armored vehicles during the rice-growing season. When it became known that Vietnam was a country with a monsoon climate and a rice culture, Americans who had been to Korea remembered the drowning summer rains that made the countryside impossible to traverse for almost half a year. Actually the Vietnam monsoons are quite different from those in Korea and do not impose the same limitations on movement. Vietnam's rice culture is, moreover, confined to a narrow belt of lowlands along the coast and the vast stretches of the Mekong Delta.

One-half of Vietnam is mountainous. Recalling the impassable mountains of some of the Pacific islands and Korea, and the extreme difficulty that armored vehicles had in operating in both places, many planners concluded that Vietnam's mountains were probably at least as rugged as those of Korea and were covered with jungle as thick as that of the Pacific islands. These assumptions were taken as additional evidence that armored vehicles had no place in Vietnam.

Yet another contribution to the growing body of notions that formed early U.S. Army attitudes toward armored units in Vietnam was a singular lack of doctrine for mounted combat in areas other than Europe and the deserts of Africa. As late as November 1961, Field Manual 17–30, The Armored Division Brigade, in a section on combat in difficult terrain, devoted one brief fourteen-line paragraph to combat in woods, swamps, and lake areas. Here it was stated that armored units should bypass, neutralize by fire, or let infantry clear difficult terrain. The basic armored tactical manual, Field Manual 17–1, Armor Operation, Small Units, devoted but six skimpy paragraphs to jungle operations.

[2] In 1954 General Westmoreland attended a Pentagon staff briefing on Vietnam that strongly influenced the senior staff officers. The Surgeon General of the Army painted a gloomy picture of an environment so severe that Westerners would be unable to survive. In July 1965 at a similar briefing when the Army Chief of Staff was present a general officer stated that tanks were not needed in Vietnam.

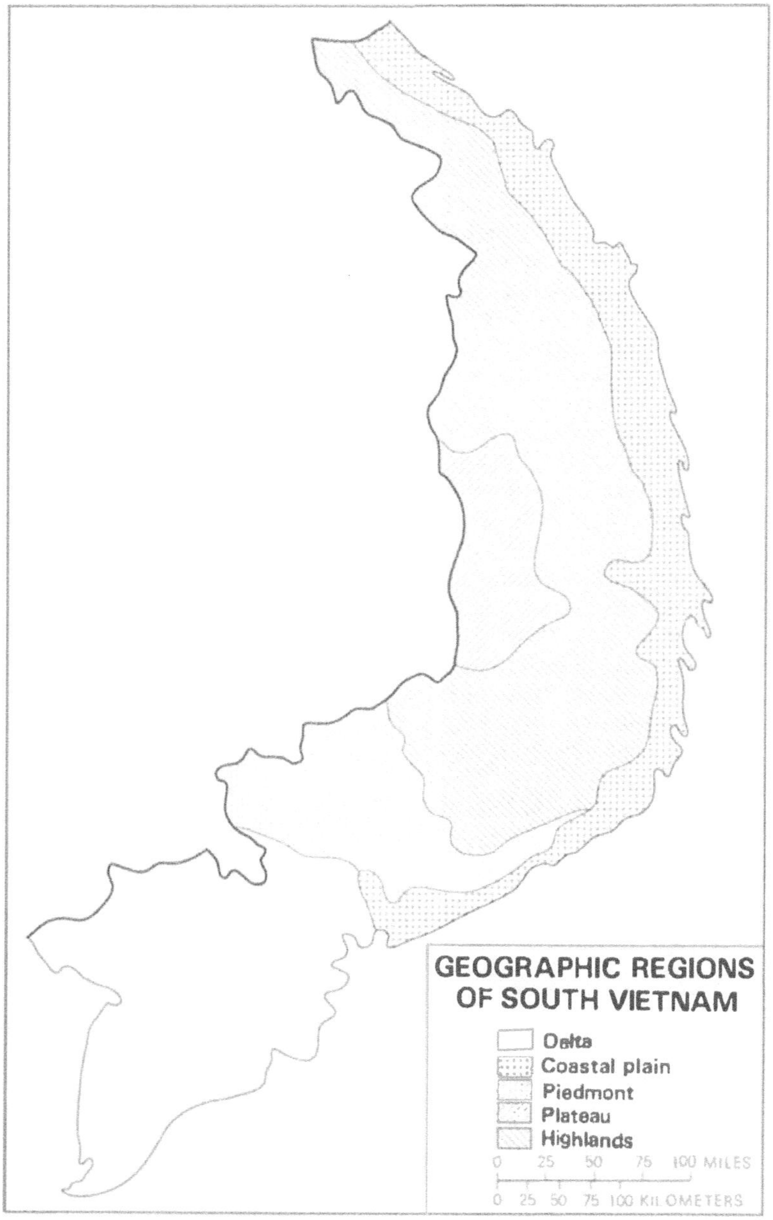
GEOGRAPHIC REGIONS
OF SOUTH VIETNAM
Delta
Coastal plain
Piedmont
Plateau
Highlands
0 25 50 75 100 MILES
0 25 50 75 100 KILOMETERS

MAP 1

Vietnam as a Field for Armor

In fact, Vietnam is not a land totally hostile to armored warfare. When the terrain was examined in detail on the ground, as it was in 1967 by a team of U.S. armor officers, it was found that over 46 percent of the country could be traversed all year round by armored vehicles. During the Vietnam War operations with armored units were conducted in every geographic area in Vietnam, the most severe restrictions being experienced in the Mekong Delta and the central highlands.

The Mekong Delta, often below sea level and rarely more than four meters above, is wet, fertile, and extensively cultivated. The area is so poorly drained that the southern tip of the country, the Ca Mau Peninsula, is an expanse of stagnant marshes and low-lying mangrove forests. Because the entire delta is criss-crossed with streams, rivers, and canals, traffic was forced to follow dikes, dams, and the few built-up roads.

In contrast to the delta, the highlands are rugged small mountains of the Annamite chain, with peaks rising to 2,600 meters. Heavily forested with tropical evergreen and bamboo, they were a difficult but not impossible obstacle for armored vehicles. Roads were poor and population centers small and scattered. When first introduced into the highlands, armored units cleared roads and escorted convoys. Subsequently, as larger enemy forces appeared, combined arms task forces operated in the mountain and jungle strongholds. *(Map 1)*

The other regions of South Vietnam—the coastal plain, piedmont, and plateau—are characterized generally by rolling or hilly terrain. Vegetation ranges from scrub growth along the coast to rice paddies, cultivated fields, and plantations through the southern piedmont, with bamboo, coniferous forests, or jungle in the northern piedmont and plateau. These areas could be used by armored ground vehicles over 80 percent of the time and were traversed by French and Vietnamese armored forces before the arrival of American troops.

The weather in Vietnam is controlled by two seasonal wind flows—the summer, or southwest, monsoon and the winter, or northeast, monsoon. The stronger of these winds, the summer monsoon, blows from June through September out of the Indian Ocean, causing the wet season in the delta, the piedmont, and most of the western highlands and plateau. The remainder of the country has its wet season from November to February during the winter monsoon, when onshore winds from the northeast shed their moisture over the northern one-third of South Vietnam.

During the transition between wet and dry periods and in the dry periods themselves, mounted combat was feasible in most parts of Vietnam. Even in the wet season, armored units proved able to operate with relative ease in many areas previously considered impassable. In 1967 a study under the title Mechanized and Armor Combat Operations, Vietnam (MACOV) was undertaken to make an extensive evaluation of the effects of Vietnam's monsoon climate on the movement of armored vehicles. Although Army engineers had conducted earlier surveys, the engineers were found to be conservative in their estimates. When there was doubt that armored units would be able to maneuver in certain places these were marked impassable by the engineers, who apparently took care that no commander would find his units stuck in an area that had been marked good for land navigation.

The group conducting the study approached the matter positively; that is, it indicated as feasible for operations any terrain where experience showed tracked vehicles had gone and could go with organic support. Applying terrain analysis data to the "go or no go" concept, the team produced maps of Vietnam for both the dry and the wet seasons. *(See Maps 2 and 3.)* More definitive and more optimistic than the engineers, the group determined that tanks could move with organic support in 61 percent of the country during the dry season and in 46 percent during the wet season. Armored personnel carriers could move in 65 percent of Vietnam the year-round.

The study confirmed what was already known to the Vietnamese: major portions of Vietnam were suitable for armored operations. But this study was not completed until almost two years after the arrival of the first Army ground combat units. During those two years many of the units were sent to Vietnam without their tanks and armored personnel carriers. Some units were even converted from mechanized infantry to infantry before deployment. The earlier studies had provided the overriding rationale for the decisions of 1965 and 1966.

The Enemy in Vietnam

By the late 1950's the insurgents in South Vietnam were known as Viet Cong, a contraction of a term that meant Vietnamese Communists. Although the enemy's methods of fighting and his ultimate goals had not really changed since the campaigns against the French, neither Vietnamese nor U.S. military observers recognized the fact. Enemy soldiers were variously described as bandits, rebels, or political malcontents; closer study revealed that the enemy was

THE DRY SEASON
SOUTH VIETNAM
No Go, Tank, APC
No Go, Tank
0 25 50 75 100 MILES
0 25 50 75 100 KILOMETERS

MAP 2

a well-organized force whose methods were the same as those of the Viet Minh against the French.

Lightly equipped and operating in a country with a primitive road network, fast-moving Viet Cong forces on foot proved more than a match for South Vietnamese troops confined to the roads. In the first stages, the Viet Cong avoided units of the Army of the Republic of Vietnam and operated as guerrillas. Sabotage, bombing, terrorism, and assassination were their hallmark. Speed, security, surprise, and deception were keys to their success.

There were many in the early 1960's who still believed the enemy was a loosely organized body with no staying power against a modern army. The truth was that the Viet Cong were well organized in regular (main) forces, provincial (local) forces, and village military (guerrilla) forces. This organization did not come about overnight; rather the Viet Cong passed through several stages that were dictated by various military and administrative situations in different parts of South Vietnam. Thus, many U.S. observers in Vietnam and the military in general did not at first realize the full extent of the enemy threat.

After 1959 small Viet Cong units began to organize into companies and battalions; guerrilla operations were a complementary tactic. Guerrilla strength grew, and secret bases were established all over the Republic of Vietnam, particularly in the lower Mekong Delta, the area north of Saigon, and the remote highlands of the north. Raids and even occasional battalion-size attacks became more frequent. These large-scale operations were centrally directed by the Lao Dong, a branch of the Communist Party of North Vietnam, through the Central Office for South Vietnam, commonly called COSVN.

An important factor in the enemy's intensification of the war was the establishment of routes for moving men and supplies from North Vietnam. Infiltration routes were in operation by 1960 and were improved and expanded during the war. Monsoon weather affected the volume of the flow and produced a pulsating effect in these arteries of men and materiel. In dry seasons and transitional periods between monsoons the flow increased dramatically, often up to four and five times the ordinary volume. The regularity of this flow in turn determined the intensity of combat that could be supported in South Vietnam. This seasonal effect of the weather would eventually be recognized in the late 1960's as a dominant factor in the enemy's scheme of operations.

Enemy supplies were limited at the beginning to relatively unsophisticated weapons and war material in limited quantities. Troops were usually former residents of South Vietnam, indoctri-

THE WET SEASON
SOUTH VIETNAM
No Go, Tank, APC
No Go, Tank
0 25 50 75 100 MILES
0 25 50 75 100 KILOMETERS

MAP 3

nated by North Vietnam as replacements for Viet Cong units. As the supply of South Vietnamese dwindled, North Vietnamese soldiers began to appear, first as replacements in Viet Cong units, then as entire units of the North Vietnamese Army. The appearance of whole units marked the transition to the last phase of the war, which was a clash between modern armies, even though Viet Cong guerrilla activities continued.

Viet Cong and North Vietnamese battle tactics invariably followed a simple formula, adopted originally from the Chinese combat doctrine of Mao Tse-tung: When the enemy advances, withdraw; when he defends, harass; when he is tired, attack; when he withdraws, pursue. To this formula was added a combat technique of "one slow, four quick," practiced with meticulous precision in almost every situation.

The first step, one slow, meant prepare slowly; a thorough and deliberate planning preceded any tactical operation. Each action was rehearsed until every leader and individual was familiar with the terrain and his specific job. Only when the commander was convinced that the rehearsal was perfect, was the operation attempted.

Execution was in four quick steps, the first of which was advance quickly. The Viet Cong moved rapidly from a relatively secure area to the objective and there moved immediately into the second step, assault quickly. In the assault, they tried to insure surprise, pouring large volumes of fire on their objective. They swiftly exploited success and pursued the enemy, killing or capturing. The third step, clear the battlefield quickly, consisted of collecting and carrying away all weapons, ammunition, and equipment, and destroying anything that could not be carried off. The Viet Cong made every effort to evacuate their wounded and dead. Finally, with orderly precision, the fourth step, withdraw quickly, was taken. The troops moved over planned withdrawal routes, with large units quickly breaking into small groups and losing themselves in as large an area as possible. Later, the scattered groups reassembled in a safe area.

Perhaps the most unusual Viet Cong fighting technique was that of carrying on a different kind of war in each of South Vietnam's forty-four provinces. South Vietnamese defenders in the northern highlands were confronted with enemy tactics that were in sharp contrast to those used in the broad southern deltas. Even more unusual was the fact that the level of conflict in each province varied surprisingly. Often one province would be simultaneously subjected to large-scale mobile attacks and guerrilla harassment, while a neighboring province was left entirely alone. This selective

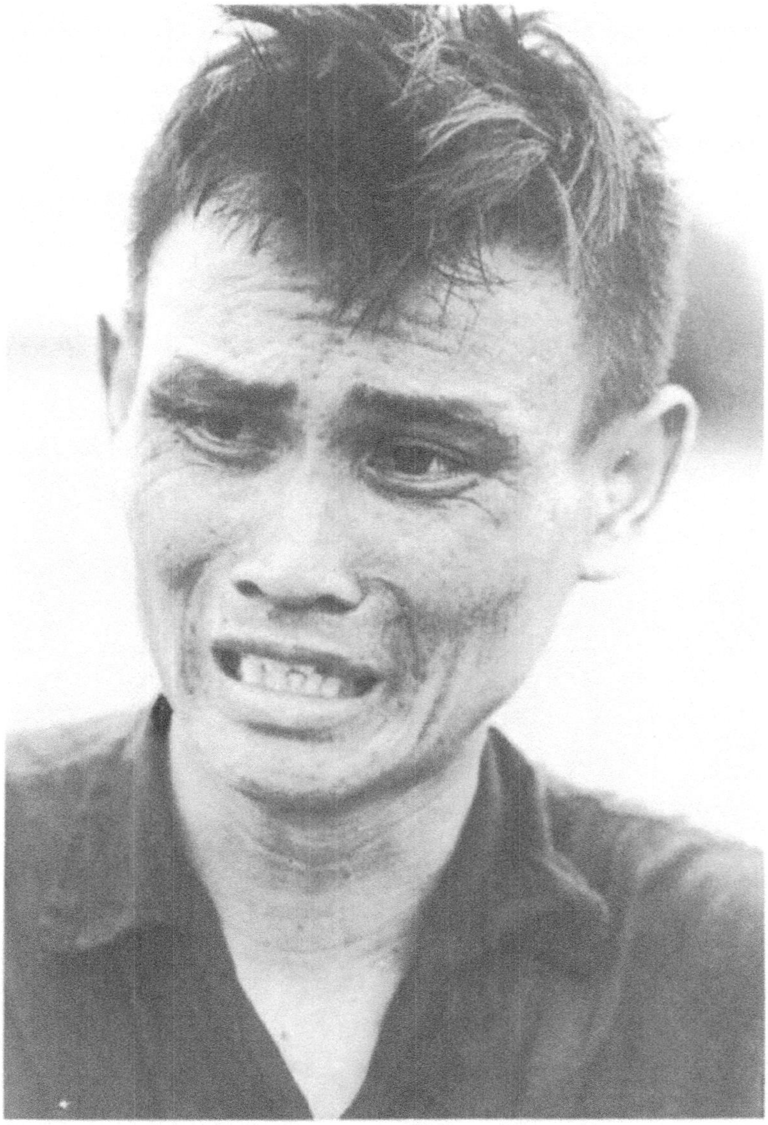

Viet Cong Soldier

intensification of the war by the Viet Cong confused American observers, and hid the true nature of the conflict. The American image of the enemy as loosely organized groups of bandits or guerrillas was not real. The enemy had a plan and worked his plan well, so well in fact that by 1964 he was ready to make the transition to the last phase of the conflict, full-scale mobile war.

CHAPTER II

Armor in the South Vietnamese Army

As modest as the armored forces were in the South Vietnamese Army, they had been in existence almost five years before the first U.S. advisers arrived. Under French tutelage, the Vietnamese Armor Corps was founded in 1950 by the simultaneous establishment of an armor training section in the Vietnamese Military Academy at Dalat and a Vietnamese reconnaissance company manned by French officers and Vietnamese enlisted men.

By late 1955, after the division of Vietnam under the Geneva Accords, the armored force in the south had an armored regiment deployed in each of the four military regions. Each regiment had a headquarters company and three reconnaissance companies that were equipped with M8 armored cars, M3 half-tracks, M3 scout cars, and towed M8 howitzers. From the French the armor troops had inherited old equipment, black berets, and concepts of employing armored units that would take years to change. Tactics stressed defense, reaction rather than action, piecemeal commitment of armored vehicles and units, the use of armored vehicles as mobile pillboxes, and the movement of armored units almost exclusively along roads.

Late in 1952 an armor course was established at the Thu Duc Reserve Officer's School. In 1955, when the officer's school became the Thu Duc Military School Center, the armor branch became a separate school. Eventually the Armor School became an independent entity under the Armor Command. The first class of the basic armor officer course was graduated in May 1955, and in June Lieutenant Colonel Walter J. Landry became the first U.S. adviser to the school. In 1956 for the first time Vietnamese officers attended the U.S. Army Armor School at Fort Knox, Kentucky, to supplement their armor training in Vietnam. Between 1956 and 1973, 712 Vietnamese officers attended courses at Fort Knox.

To oversee armor training and the development of armor tactics for the South Vietnamese Army, the Vietnamese Armor Command, which also served as the office of the Chief of Armor, was established on 1 April 1955. The Chief of Armor, a general staff officer, theoretically had no tactical authority but advised the Vietnamese high command on armor matters, particularly training. He also supervised the armor school, armor doctrine, force struc-

SOUTH VIETNAMESE ARMORED RECONNAISSANCE UNIT *stands inspection in 1952 at Thu Duc Officers School. The vehicle is a 1939 French Panhard armored car.*

ture, and equipment changes. In the early years the Chief of Armor became very powerful and even gave orders to field units, but this abuse of the chain of command ceased in the mid-1960's.

With the creation of the Republic of Vietnam in October 1955 the Vietnamese Armor Corps became a part of the Army of the Republic of Vietnam, and by early 1956, when the U.S. advisers appeared it was already fairly well organized both for training and operations in the field. The existing South Vietnamese armored units were reorganized under U.S. influence as armored cavalry regiments consisting of two reconnaissance squadrons, each equipped with M8 armored cars, M3 half-tracks and M3 scout cars, and one squadron of M24 tanks.[1] The Armor School, swept along in the reorganization, was patterned after the U.S. Army Armor School at Fort Knox. Instruction emphasized the hands-on method that characterized the American school, lesson plans and manuals were adapted from those prepared at Fort Knox, and courses like those at Fort Knox were undertaken to cover the full range of armor training.

[1] A Vietnamese regiment was equivalent in size to an American battalion or squadron, and a Vietnamese squadron was equivalent to an American company or troop.

In the years from 1957 to 1959 the U.S. adviser group with the armored force was small; sixteen armor officers were assigned to the Military Assistance Advisory Group, Vietnam. Only four armor officers were assigned to armored field units, and these assignments were not made until 1959. Not only were field advisers restricted from participating in combat operations, but before 1962 all returning military advisory officers were obliged to sign a certificate binding them to secrecy on the subject of advisory operations in Vietnam. This policy limited discussion and the publication of articles and hampered the U.S. Army in its own preparations for Vietnam. It is no wonder then, considering the little armor there was in South Vietnam, the few American advisers, and the seal of secrecy, that the experiences of Vietnamese armored units were ignored by the U.S. military, even during U.S. troop deployments in 1965 and 1966. Yet this experience was gained in the hardest school of all, combat against an enemy. Ultimately, the impression of Vietnamese armored forces that filtered through the screen of secrecy was one of a force that had played a minor role in warfare in Vietnam. True or not, the impression was created, at least in part by the U.S. advisers.

U.S. Advisers

In early 1962, with the formation of the U.S. Military Assistance Command, Vietnam (MACV), the number of American advisers grew. In the fall of 1962, ten more armor officers arrived: four replaced armored cavalry regimental advisers, the others were assigned to six newly activated squadrons. The armor advisers who came to teach Vietnamese armor commanders and soldiers also learned, largely through trial and error, about the land, the war, and the enemy. By 1965 the regimental advisory detachment was authorized a major as regimental adviser, a captain as staff adviser, and a company grade officer and two noncommissioned officers for each squadron. As the armored force expanded, each new regiment and squadron was assigned an advisory detachment. These detachments preceded U.S. units into Vietnam and remained behind them until January 1973.

Advisers formed an exclusive group of officers and noncommissioned officers who were chosen by the Department of the Army strictly because they were available for reassignment and advisers were needed in Vietnam. There was no special schooling for advisers until early 1962 when a six-week course was established at Fort Bragg, North Carolina. This advisers' course, under the auspices of the U.S. Army Special Forces School, was basically infantry oriented and no armor training was provided. Language instruction was not included and was offered only at formal language schools. As a result, most advisers had no knowledge of the Vietnamese language—a deficiency that continued as long as the war lasted. No adviser was really prepared for the cultural, climatic, and professional shock that awaited him when he joined a South Vietnamese Army unit. Patience, perseverance, and common sense were the most important assets a man could have for the job.

Once in Saigon, armor advisers were assigned to Vietnamese units by the armor adviser on the staff of the Military Assistance Advisory Group. Later, these assignments were made by the senior U.S. adviser to the South Vietnamese Armor Command. Each new adviser was given a short briefing on South Vietnamese armor and on the unit he was to advise before he left by helicopter or jeep to join his unit. Because squadrons were widely separated until mid-1966, the adviser rarely saw his boss, the regimental adviser, and seldom operated with other squadrons of the regiment. Dependent for supplies, food, ammunition, and his very existence on his South Vietnamese unit, he had to establish himself quickly as a team member.

The duty of an adviser can best be described as frustrating; it was complicated by serious differences in culture and combat experience between adviser and advised and the fact that the adviser could not speak the language of those he advised. As a representative of the U.S. Army, the adviser had access to communication and firepower support vastly superior to that of the Vietnamese unit commander. Although this advantage sometimes helped to establish the necessary working relationship, the adviser achieved a good relationship by working all day, seven days a week, mostly in the company of the unit commander. Although not expected to fight, on operations the adviser went with the Vietnamese commander, providing advice and American assistance with fire support, communications, and medical evacuation.

Until mid-1965 advisers were the main source of information for the American military on armor operations in Vietnam. When American troops finally joined the fighting, the adviser's position often became more difficult, particularly in the early years of the American troop buildup when the Army of the Republic of Vietnam was generally regarded by U.S. units as inferior and a security risk. Frequently, the adviser himself was also treated like a poor country cousin. Many advisers declared bitterly that American armored units often ignored the experience and knowledge of enemy and terrain offered by the Vietnamese armor officers and the advisers. As a result of this attitude American units made frequent mistakes that Vietnamese armor avoided as a matter of course. Eventually, time, improvement in the Vietnamese Army, and repeated tours by U.S. soldiers altered the situation. In the later stages of the war, many Vietnamese and American armored units established joint training, operational, and social programs that were helpful to units of both armed forces.

Most armor advisers were impressed by the technical proficiency of their Vietnamese units. In the maintenance of weapons and equipment, the Vietnamese armor crewman was outstanding; his innovations often extended to making repairs to his equipment with chewing gum, bailing wire, and even banana stalks. In tactics and the coordination of fire support, South Vietnamese armored units had much to learn in the early years from American advisers. The exchange of information between adviser and unit gave the adviser a better understanding of Vietnam's particular problems and was indirectly useful in helping to prepare American units for service in Vietnam.

Perhaps the contribution of armor advisers can best be summarized by saying that they created an atmosphere in which change, innovation, and development could take place. They were in a sense responsible for the spirit of aggressiveness and confidence that grew in Vietnamese armored units, and eventually led to the successes of Vietnamese armor during *Tet* 1968 and the enemy offensive of 1972.

M113's in the Mekong Delta

Despite the efforts of U.S. advisers, South Vietnamese Army tactics remained French for some time and reflected Vietnamese political pressures that put a premium on holding down casualties in men and equipment. Armor was used principally in defense; its primary missions were escorting convoys, defending positions, clearing roads, and acting as relief forces. Vietnamese equipment had been used in the French war and age had aggravated its shortcomings.

In an effort to introduce more recent equipment, in late 1961 the Military Assistance Advisory Group studied the possibility of using the M113 armored personnel carrier (APC) in Vietnamese Army units in the Mekong Delta. On the basis of recommendations growing out of this study, two M113 companies were organized in April 1962 and assigned to the Vietnamese 7th and 21st Infantry Divisions. The original plan had been to issue the armored personnel carriers to two well-trained rifle companies, then instruct the troops in mechanized operations. The newly organized units were not well trained; one company was a hodgepodge of troops selected at random. Most key command positions were filled with men from the South Vietnamese Armor Command who had little combat experience, and it was necessary to extend the six-week training program to nine weeks. The companies were joined by U.S. advisers Captain James W. Bricker and Captain Stanley E. Holtom, recent graduates of the first advisers' course at Fort Bragg. Originally called mechanized rifle companies, the new units were modeled after American mechanized rifle companies. Each company was organized into three rifle platoons with three APC's in each platoon; a support platoon with four APC's, three 60-mm. mortars, and three 3.5-inch rocket launchers; and a company headquarters section with two APC's. All carriers were equipped with a .50-caliber machine gun, and eighteen .30-caliber Browning automatic rifles were distributed throughout the company.

As with almost any new organization, the units' first engagements reflected their lack of experience. Because of the limited knowledge of armored tactics at higher commands, the units were also handicapped with improper assignments or missions that led them into impassable terrain. These early operations provoked a great deal of unfavorable comment that was directed chiefly at the alleged inadequacies of the M113. The situation improved, however, when the units acquired experience. Between 11 June and 30 September 1962, the two companies killed 502 of the Viet Cong and took 184 prisoners at a cost to themselves of 4 dead and 9 wounded.

The first real success with the new mechanized forces came about in a Vietnamese 7th Infantry Division operation in September 1962 in the Plain of Reeds. After studying the South Vietnamese plan, which involved several regiments under division control, Captain Bricker, the mechanized unit adviser, recommended that because of the terrain the mechanized company not be used. But since the division commander had personally developed the scheme of maneuver no changes were made and the operation was launched as scheduled.

Although aerial reconnaissance had confirmed the terrain difficulties, the mechanized company moved out early on 25 September with nine armored personnel carriers. For a time all went smoothly and movement was rapid. The company was ordered to bypass its first objective, continue to the second, and block the Viet Cong escape route along a four-kilometer segment of a canal. Shortly thereafter it was ordered to cross the canal and attack a group of sixty Viet Cong.

ARMORED PERSONNEL CARRIER, M113, *firing .50-caliber machine gun during South Vietnamese training exercise. Barrel of side-mounted .30-caliber machine gun can be seen on far side of M113.*

The company reached the canal about 1045 and the lead elements, after a delay in finding a suitable site, began to cross. When one platoon had reached the other side, an observation post spotted fourteen enemy soldiers and the South Vietnamese company commander, Captain Ly Tong Ba, decided to attack. Although Captain Bricker had urged that the armored carriers be used to maneuver to the enemy flank, the APC's charged straight ahead through the flooded paddies toward the point where the Viet Cong had last been seen. Suddenly, enemy soldiers appeared all around the APC's, some firing automatic weapons and rifles and others running wildly in an attempt to evade the armored vehicles. As the APC's scattered the enemy, the South Vietnamese soldiers fired in all directions from the open hatches, with the .50-caliber machine guns dominating the fight.

After much discussion, Captain Bricker persuaded Captain Ba to order his troops to dismount, but this move proved to be a serious mistake. As long as the troops were mounted and moving the Viet Cong had been unable to fire well-aimed shots. Dismounted, the soldiers not only lost the advantages of movement, cover, and observation afforded by their APC's but they also found themselves bogged down in water that was knee-deep. No sooner were the armor troops immobilized than accurate enemy fire found them out; the Viet Cong, knowing the terrain, had moved to the higher ground of an underwater knoll. As squad control disintegrated, casualties increased and the operation became disorganized.

The fighting subsided after an hour and South Vietnamese troops remounted their vehicles and moved west. Almost at once the company was engaged by heavy small arms fire from isolated groups of the enemy. As the APC's attacked, enemy soldiers hiding in the reeds were flushed from their concealment and overrun. It was midafternoon before the enemy began to withdraw. When South Vietnamese troops cleared the battlefield they found more than 150 of the Viet Cong dead. They had also captured thirty-eight men and seized twenty-seven weapons, including an American .50-caliber machine gun and two Browning automatic rifles.

Despite difficult terrain and poor tactical decisions this mechanized operation achieved success. Generally the unit performed well, showing that it had made great progress since completing its training in June. Its failure to maneuver, however, prevented the rapid destruction of the enemy, and, combined with confusion as to whether to conduct the attack mounted or dismounted, demonstrated that both the Vietnamese armor soldier and his adviser still had much to learn. The application of what was then American doctrine, which called for mechanized troops to dismount and assault the objective on foot, caused the attack to falter. Because of this experience, attacks in the future were conducted with troops fighting from the carriers. The M113 quickly became a combat vehicle, used almost as a light tank. Eventually, American units adopted this doctrine and fought mounted.[2]

Reorganization and Retrenchment

The success of the first two South Vietnamese mechanized companies demonstrated the value of highly maneuverable, lightly armored vehicles in Vietnam. To supplement these two companies, the newly developed M114 reconnaissance vehicle and additional M113's were shipped to Vietnam. The first two M113 companies were redesignated the 4th and 5th Mechanized Rifle Squadrons of the 2d Armored Cavalry Regiment and assigned to the IV Corps Tactical Zone in the Mekong Delta. The armored cavalry regiments

[2] The appearance of the green camouflaged M113's, moving rapidly over the water-soaked fields, belching fire and smoke from their machine guns and engines, gave rise to the nickname "green dragons."

supporting each of the four tactical zones were reorganized in late 1962 with the addition of one armored reconnaissance squadron and two mechanized rifle squadrons. M113's for the new squadrons arrived in late 1962 and the squadrons became operational as they completed training at the South Vietnamese Armor School. By May 1963 each of the four regiments had one squadron each of M24 tanks, M8 armored cars, and M114's, and two mechanized rifle squadrons with M113's.

Mechanized rifle squadrons were organized like their predecessors, the mechanized rifle companies, except for the supporting weapons. During 1962 the 3.5-inch rocket launcher and the 60-mm. mortar were judged unsatisfactory because of their limited range. Each newly organized squadron was equipped therefore with three 81-mm. mortars and a 57-mm. recoilless rifle, all transported by armored personnel carriers.

Unit reorganization and new equipment alone were not enough to bring about a change in the war. New tactics, better leadership, and improved training were needed to complement the increased firepower. When training lagged, overconfidence and poor leadership combined to teach some costly lessons. Advisers were often frustrated in their attempts to persuade Vietnamese armor leaders to use the new APC's properly and aggressively.

The battle of Ap Bac, sixty-five kilometers southwest of Saigon in the Mekong Delta in January 1963, illustrates the problem. The month before, Vietnamese Army intelligence had reported a reinforced Viet Cong company in Ap Tan Thoi, 1,500 meters northwest of Ap Bac. *(Map 4)* The Vietnamese 7th Division planned an operation to trap the Viet Cong by landing the 11th Infantry Regiment to the north by helicopter while a provisional regiment of two battalion-size task forces of Civil Guards (later named Regional Forces) moved in from the south. The 4th Mechanized Rifle Squadron, 2d Armored Cavalry, commanded by Captain Ba, was attached to the provisional regiment and was to attack from the southwest. Three Vietnamese Ranger and infantry companies were in reserve, with artillery and air support on call.

In contrast to the intelligence estimate, the enemy force actually consisted of three main force companies, reinforced with machine guns, 60-mm. mortars, and several local guerrilla units. The Viet Cong after action report subsequently captured revealed that the enemy knew a battle was imminent and had carefully prepared defensive positions along the Cong Luong Canal from Ap Tan Thoi to Ap Bac. The canal, bordered with vegetation, offered concealment and unobstructed fields of fire across the open rice paddies.

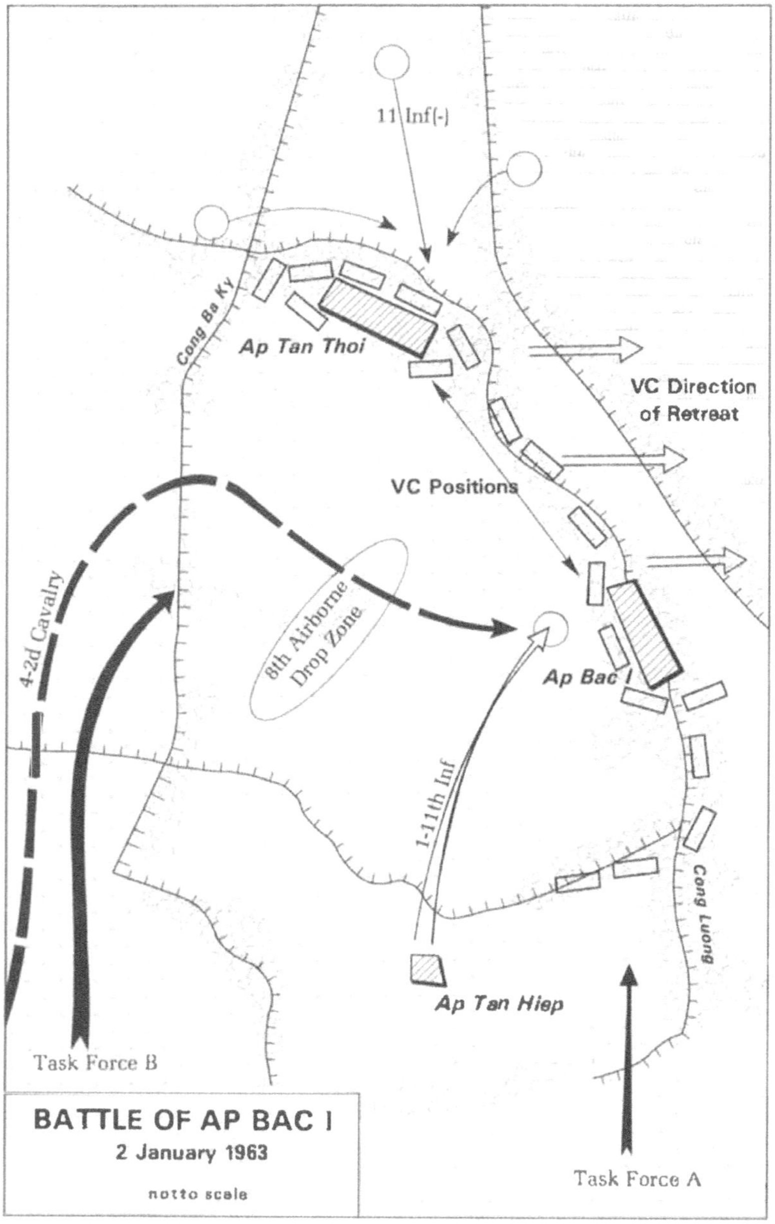
11 Inf(-)
Cong Ba Ky
Ap Tan Thoi
VC Direction of Retreat
VC Positions
4-2d Cavalry
8th Airborne Drop Zone
Ap Bac I
1-11th Inf
Cong Luong
Ap Tan Hiep
Task Force B
Task Force A
BATTLE OF AP BAC I
2 January 1963
notto scale

MAP 4

On the morning of 2 January 1963 the Civil Guard task forces started north, while in three uneventful trips helicopters lifted the Vietnamese 11th Infantry Regiment into position. About 0730 Task Force A encountered the southern flank of the Viet Cong positions along the Cong Luong Canal. During the first moments of battle, the task force commander was wounded and a company commander killed. Major Lam Quang Tho, commander of the 2d Armored Cavalry Regiment and also province chief, refused to allow the provincial forces to advance, and changed their mission to one of occupying blocking positions. Colonel Bui Dinh Dam, 7th Division commander, decided to commit a reserve force to the west side of the canal that runs through Ap Bac. At 1020 as the helicopters came in for their fourth lift, the Viet Cong antiaircraft crews hidden along the canal opened fire. Of the 15 helicopters bringing in the reserve, 14 were hit, and by noon 5 had been shot down.

Lieutenant Colonel John P. Vann, division adviser, radioed Captain James B. Scanlon, senior adviser to the 2d Armored Cavalry Regiment, that the helicopters were down about 1,500 meters to the southeast of the regiment. After considerable argument with Captain Scanlon, Captain Ba finally agreed to move across the Cong Ba Ky Canal and secure the helicopters. Three hours later, as the first of the APC's approached the helicopters, enemy fire suddenly raked the two leading vehicles and their dismounted infantrymen. The APC's began backing up, abandoning the wounded. A few minutes later they advanced again, firing their .50-caliber machine guns, and again they were hit by enemy fire. Exposed from the waist up, the machine gunners were particularly vulnerable; fourteen of them died before the day was over.

Captain Scanlon ran to the aid of the wounded helicopter crews, and with the help of Sergeant Bowers, another adviser, carried them to the APC's. By now more APC's had crossed the canal and they too tried to maneuver forward, but without success. Because there was no unified effort, the Viet Cong was able to concentrate fire on each vehicle in turn as it ventured forward.

By midafternoon when it was apparent that the enemy could not be overrun, the South Vietnamese Army commanders and the advisers decided to request reinforcement by an airborne battalion. Despite the vehement objections of both Colonel Vann and Colonel Daniel B. Porter, who was the IV Corps Tactical Zone adviser, the corps commander decided to drop the South Vietnamese airborne battalion to the west, behind the mechanized squadron, rather than east of the canal, where it would have completed the encirclement. At dusk the 8th Airborne Battalion parachuted into the rice paddies. The night was quiet save for artillery fire and the pop of flares over the enemy positions. Taking advantage of the open eastern side, the Viet Cong withdrew during the night and were gone by daylight. Early in the morning dismounted troops from the APC squadron crossed the canal, passed the empty enemy positions, and swept through most of Ap Bac before being ordered to hold. The airborne battalion was still organizing and collecting parachutes and was not ready to attack. Finally, at noon the force staged an attack that was really nothing more than a walk-through.

Because of the large number of South Vietnamese troops involved, and especially because of the number of U.S. helicopters downed early on the first day, the battle of Ap Bac drew much attention. Although estimates of the results of Ap Bac varied, Colonel Vann considered the operation a failure. Several days later he stated: "There were three main criticisms. . . . First, the failure of South Force to move. Second, failure of the APC's to move. Third, the parachute force. They were dropped on the wrong side of the river. It was a decision that I opposed. They wanted to reinforce defeat rather than ensure victory."

The fighting at Ap Bac and more specifically the employment of the mechanized rifle squadron, illustrates many of the problems faced by advisers. Poor coordination and planning were apparent at all levels in the South Vietnamese command: the airborne forces were not correctly employed and there was no unity of command on the ground. Politics also played a part. The South Vietnamese cavalry commander was the political leader of the province, and because his political and military future depended on his keeping casualties in the Civil Guard and armored cavalry to a minimum, he was reluctant to have these forces attack. Finally, strained relations between the advisers and the South Vietnamese unit commanders materially contributed to the lack of cohesion. Politically, the battle was reported as a victory, but for the armored forces it was much less. Only when the crews had had further training and experience and when improvements had been made on the equipment would the APC's be employed to better advantage.

While the lessons of Ap Bac were still fresh, the South Vietnamese armored force undertook a series of expansions. In December 1963 two regimental-size armored units were activated. The 5th Armor Group, later redesignated the 5th Armored Cavalry Regiment, and the 6th Mechanized Battle Group, later redesignated the 6th Armored Cavalry Regiment, were both assigned to the Vietnamese high command as a general reserve. Formed from the tank squadrons of other units, the 5th Armor Group was in effect the first South Vietnamese Army tank regiment. Later, in March 1964, two more mechanized rifle squadrons were formed at the Armor School and completed their training in October. These additions made a total of fourteen operational mechanized rifle squadrons.

Even with additional forces and equipment, however, the combat record of South Vietnamese armored forces in 1964 was still uneven. One battle in the Plain of Reeds on 3 and 4 March 1964 ended in a resounding victory for the South Vietnamese and the capture of over 300 of the Viet Cong. In contrast, on 28 December 1964 the 9th Viet Cong Division seized the town of Binh Gia, sixty-five kilometers east of Saigon. During a battle that lasted several days, South Vietnamese Ranger and Marine battalions were severely beaten. As the fighting continued into the first few days of 1965, South Vietnamese armored relief forces were ambushed and they too suffered heavy casualties. This battle was significant to both sides since it marked the beginning of Mao Tse-tung's classic and final "mobile" phase of the war. For the first time a Viet Cong division was committed as a whole: the battle itself was unusual in that the enemy unit remained on the battlefield for several days instead of resorting to the customary Viet Cong hit-and-run tactics.

Almost at the same time that the Viet Cong began to appear in larger units, American forces began deploying in Vietnam. Probably the most publicized campaign of 1965 was that of the U.S. 1st Cavalry Division in the Ia Drang valley west of Pleiku in the II Corps Tactical Zone. Little publicity, however, was given to the actions which led up to the 1st Cavalry battles in the Ia Drang valley.

From 19 to 27 October a South Vietnamese armored task force of tanks and APC's with Rangers battled through to the relief of the Plei Me Special Forces camp southwest of Pleiku. Although ambushed en route by a North Vietnamese regiment, the task force, with U.S. air and artillery support, reached the camp, established a perimeter, and stood off a heavy attack. It then counterattacked, killing 148 of the enemy and capturing 5. It was during this campaign that enough intelligence information was collected to warrant sending the 1st Cavalry Division into the Ia Drang valley.

The fight to save the Plei Me Special Forces Camp allowed American advisers to observe the Vietnamese armored task force in action. Gunnery was poor in the tank units, which were now equipped with M41 tanks. There was no effort to place accurate main gun fire; instead, the Vietnamese tankers "shot from the hip" in the enemy's general direction and kept firing as fast as possible. Tanks bunched up, and troops demonstrated little aggressiveness, content to stand and fight as if they were in pillboxes. Security was poor—the unit was ambushed on the very first day. Coordinated action between tanks, armored personnel carriers, and Rangers was almost nonexistent. Leadership and control were still a long way from acceptable standards. Maintenance continued to be a bright spot in the performance of Vietnamese armored troops; all fifteen tanks returned from the fight.

Expansion of Armor in the South Vietnamese Army

In February 1965 Major Lloyd J. Brown, training adviser to the Vietnamese Armor Command proposed that the armored force be increased by seven mechanized rifle squadrons, and that the first squadron begin training almost immediately. A parallel plan sought to increase manpower authorizations and thus provide one armored cavalry regiment for each Vietnamese division, retain a task force headquarters, and greatly improve the efficiency of administrative and logistical support to the squadrons. Subsequently approved by the Vietnamese Joint General Staff and completed by mid-1966, the expansion helped solve some of the problems caused by wide dispersion of units.

Reorganization of South Vietnamese armored units in mid-1965 brought about a change in the U.S. advisory organization. In the early 1960's there was no advisory detachment at the Armor Command; an officer assigned to the organization and training division of the Military Assistance Advisory Group also advised the Vietnamese Chief of Armor. At first described as senior adviser to the Armor Command, he became training adviser when the advisory group was supplanted by the Military Assistance Command, Vietnam, and the position came under the Training Directorate of that command. In July 1965 when the Military Assistance Command, Vietnam, established an Armor Command Advisory Detachment, Major Brown, the training adviser, became the senior adviser of the Armor Command.[3]

Plans for the Vietnamese armored force for 1965 called for one V-100 armored car squadron and ten separate armored car troops to be formed, trained, and deployed to replace units equipped with the M8 and older obsolete vehicles. Three of these troops completed training in May, but structural flaws found in the V-100

[3] Major Brown was the first American awarded the coveted Army of the Republic of Vietnam Armor Badge, appropriately numbered 1.

armored cars delayed their use for six months. The faults were corrected by the Vietnamese Army 80th Ordnance Depot. When these units assumed route security missions the M113 units would be free for expanded combat operations.

In the midst of the expansion, the Armor Command was exploring alternatives for further increases. In late March 1966 Lieutenant Colonel Raymond Battreall, senior adviser to the Armor Command, directed Major James Madole, G-3 adviser, to study ways of enlarging the armored force. After accompanying each of the ten Vietnamese cavalry regiments on combat operations, Major Madole recommended that armored regiments be increased from ten to sixteen. Two of the new regiments would be placed in the two divisions scheduled to be added to the Vietnamese Army. Each of the other four would be placed under control of a corps tactical zone commander as a mobile reserve. Although the recommendations were not adopted at the time, the study influenced subsequent decisions to enlarge the Vietnamese armored forces.

A parallel study of Vietnamese armored cavalry was published in February 1966 by the U.S. Army Concept Team in Vietnam. This study evaluated organization, equipment, support needs, and the best ways to use armored units. After observing the six existing cavalry regiments from February to May 1965, the authors of the study recommended that each division be assigned an armored cavalry regiment consisting of a headquarters and at least two mechanized rifle squadrons. Commenting on the study, the director of the Joint Research and Test Activity in Vietnam, Brigadier General John K. Boles, Jr., agreed that additional armor units were needed. However, he considered the need to be so urgent that he suggested the deployment of U.S. or other free world armored units. Thus, indirectly, Vietnamese armor influenced the deployment of U.S. armor.

Cuu Long 15

All Free World Military Assistance Forces in Vietnam faced the problem of forcing the Viet Cong into an engagement at a time of the allies' choosing. The allies' well-laid plans, based on what appeared to be sound intelligence reports, frequently led to so-called walks in the sun—because the enemy managed to avoid combat. The following is an account of one attempt to overcome this problem.

In March 1966 Vietnamese Army forces planned to conduct a search and destroy operation near Moc Hoa in the delta using the 6th Armored Cavalry Regiment and infantry elements of the 7th

Infantry Division. To reach the Moc Hoa area and be in position to attack on 21 March without alerting the enemy, armored forces moved a day earlier to Ap Bac, twenty-five kilometers south of Moc Hoa. Reaching Ap Bac on the afternoon of 20 March 1966, the 6th Armored Cavalry Regiment moved into a previously used assembly area, resupplied, and ostensibly prepared to stay the night. To an observer, the cavalry action resembled any number of earlier operations that had swept eastward from Ap Bac along the major canal net to Tuyen Nhon. Under the cover of darkness, however, the 6th left Ap Bac and marched to Moc Hoa, arriving there at dawn.

At first light on 21 March the 6th Armored Cavalry Regiment and supporting infantry elements moved across the major canal north of Moc Hoa and began a sweep to the northeast. Simultaneously, Vietnamese infantry elements were lifted by helicopter to blocking positions between the canal and the Cambodian border. Once across the canal, the regiment deployed its squadrons on line, with the lead troop of each formation in a wedge formation and the two following troops in column. At 0900 the 1st Squadron, commanded by a Captain Tien encountered a few Viet Cong eight kilometers northeast of Moc Hoa. Captain Tien immediately deployed his squadron on line and, firing rapidly, his men moved toward the enemy. High grass, however, limited visibility and the Viet Cong slipped away. At about noon the squadron began receiving small arms fire as it moved along a canal fourteen kilometers east of Moc Hoa. Again the squadron deployed on line and pressed rapidly toward the enemy positions. Pushing their way through the tall grass, the cavalry attempted to force the enemy into the open with machine gun fire. This time, however, the enemy had moved to well-prepared fighting emplacements from which they fought back with recoilless rifle, mortar, and small arms fire.

The high grass provided the enemy with good concealment and hampered coordination among the Vietnamese units. The 1st Squadron, 6th Cavalry Regiment, and the supporting infantry mounted on armored personnel carriers found themselves heavily engaged, flanked on both sides, and receiving fire from all directions. One APC was struck on the gunshield by recoilless rifle fire that killed the gunner and wounded three U.S. advisers riding in the vehicle. The cavalry laid down a barrage and withdrew to regroup and evacuate the wounded. While Captain Jerrell E. Hamby, the U.S. adviser, directed the medical helicopters, the troopers set fire to the tall grass, burning off the vegetation and exposing the enemy fighting positions.

Again the cavalrymen and their supporting dismounted infantry deployed on line and attacked; this time with the grass burned away they were able to pinpoint and destroy the enemy with machine gun fire. A running gun battle lasted until nightfall as the combined force moved slowly eastward along the canal. More than 200 of the enemy were killed, 17 were taken prisoner, and 89 weapons were captured. The operation demonstrated that with proper planning an armored unit could operate effectively as a strike force, even against elusive insurgents. In this instance a combined arms force, aggressively led, had proved to have sufficient organic combat power to overcome a well-armed and well-entrenched foe.

Time for Corrective Analysis

Actions like Cuu Long 15 were the exception, however, for the misuse of armor continued to be a problem through 1967. Many senior Vietnamese commanders either ignored or did not understand the capabilities of an armored force. Part of the misuse was due to the complex Vietnamese concept of unit of command. In part it also reflected the persistent palace guard syndrome that caused Vietnamese commanders to parcel out their armored forces among several province, district, or other headquarters as security elements. Consequently, armored units were unable either to take the fight to the enemy or to maintain an acceptable state of combat readiness. For example, at Ban Me Thuot in II Corps Tactical Zone the 1st Squadron (tank), 8th Armored Cavalry Regiment, served as the province headquarters guard force in 1967. The squadron had not fought since early 1966. Soon after assuming his duties, the unit's senior adviser, Captain Joseph Snow, discovered that over 60 percent of the troops had never been in combat. An extensive crew training program was immediately begun.

Although not all South Vietnamese armored squadrons were as inexperienced as the 1st Squadron, 8th Cavalry Regiment, similar situations were frequently encountered by other U.S. advisers as they tried to improve unit performance. A detailed discussion of the problems involved in the use of armor took place during an armor advisers' conference in March 1967. The advisers reviewed events of the past year and presented their consolidated recommendations to the Vietnamese Chief of Armor. Subjects covered the entire range of armor employment and included doctrine, training, organization, equipment, weapons, and combat vehicles.

The advisers' recommendations strongly echoed those set forth in the 1967 U.S. Army study, and included the suggestions that

1. Army of the Republic of Vietnam Field Manual 3-1, Armor Operations in Vietnam, be approved and distributed to all commanders, service schools, and training centers:

2. An M113 hydraulically operated vehicle-launched bridge, an M113 dozer blade kit, and the M125A1 81-mm. mortar carrier be added to the Vietnamese armor's table of organization and equipment;

3. Modern weapons including the M16 rifle be issued and an automatic 40-mm. grenade launcher be evaluated;

4. Infrared equipment (M8 binoculars and xenon searchlights) be authorized and training conducted to improve night operations and the use of night sighting devices.

The advisers also supported an Armor Command proposal that the number of M113's per troop be increased from three to five, and the remaining M8 armored car squadrons be replaced with M113 squadrons.

With most formal unit training and the acquisition of equipment completed and with a temporary freeze on force levels imposed, advisers began to put more stress on improving the quality of operations. They determined that there were several reasons for the poor use of Vietnamese armored forces. Supervision of subordinate elements by regimental headquarters, for example, was difficult because squadrons were widely dispersed. In addition, the tendency to employ the squadrons piecemeal frequently left regimental headquarters as unemployed control elements during combat operations. Inexperienced senior commanders often assigned unsuitable missions to armored units. Armored commanders were often given only the barest instructions for an operation—an objective and a direction of attack. Too many units were poorly led by ineffective but politically influential commanders.

Advisers were finally obliged to bring pressure against the Vietnamese high command to improve leadership. In April 1967, for example, they suggested that U.S. equipment support funds be withdrawn from the 5th Armored Cavalry Regiment, and in August another cavalry regiment was singled out for the same purpose. The South Vietnamese high command was moved to take corrective action. As a follow-up, advisers at all levels continued to emphasize the proper use of armored forces, including the exercise of unity of command and aggressive leadership.

In operations the South Vietnamese armored units continued to improve upon existing doctrinal guidelines. When ambushed, Vietnamese cavalrymen always attacked the ambushers directly. South Vietnamese operations were always much shorter in duration than U.S. operations and the M113 units always had an infantry unit attached to fight on foot if necessary. A squadron usually operated on one radio channel so that all vehicle commanders were in constant communication and could hear the unit leader. Combat orders were oral and frequently modified to suit the developing situation. The line formation was the favorite method of movement, although enemy mining later changed this tactic. Superstition even modified doctrine since Vietnamese units seldom took eleven or thirteen APC's on an operation because these were considered unlucky numbers.

In the summer of 1967 General Creighton W. Abrams, deputy commander of the Military Assistance Command, Vietnam, approved the U.S. funding for several proposals made on behalf of the South Vietnamese high command by the Armor Command senior adviser, Colonel Battreall. Foremost was a plan to increase the number of M113's in a Vietnamese cavalry troop (platoon) from 3 to 5, and the total number per squadron (troop) from 15 to 22 without increasing the number of soldiers. Before this time APC squadrons had been organized along the lines of U.S. mechanized infantry companies, even though Vietnamese squadron tactics as they developed were not those of an infantry unit, but rather those of a tank company in mounted combat. The increase in the number of vehicles was accompanied by cutting the number of crewmen to a vehicle to seven men. Crew positions thus saved were then used to fill crews on additional vehicles. Among other approved proposals was one to disband armored car squadrons and issue M125A1 81-mm. mortar carriers. Regimental headquarters retained their armored car platoons, which had been issued V-100 armored cars in 1967. The M125A1 mortar carrier replaced the field expedient of mounting 81-mm. mortars on the M113.

The advisers' emphasis on training slowly began to produce results. The 4th Armored Cavalry Regiment in I Corps Tactical Zone was an example of a unit which performed well in combat. The 4th developed an effective technique for occupying and defending a night position. Preferring to engage the enemy in the open rather than in fortified positions, the cavalrymen set up the night location whenever possible in a large open area. Vehicles moved into the area at dusk and occupied positions separated from each other by thirty-five to forty meters, the bursting radius of an enemy mortar round. The squadron was usually reinforced with an infantry battalion, dug in on line with the front of the vehicles. Barbed wire or claymore mines were placed only on wooded approaches, leaving the rest of the area free for vehicles to maneuver during the defense. The short range of the enemy mortar and the distinctive thump produced by firing enabled the squadron to determine a compass azimuth to the mortar position. Fire from 81-mm. mortars was then walked up and down that azimuth, usually with good effect. If the perimeter received fire from long-range direct fire weapons, the squadron employed artillery and mortars rather than fire from the APC's, thus concealing the vehicles' positions.

To counter a ground attack, the tracked vehicles waited until the enemy was in the open before turning on their headlights and making a mounted assault with all weapons firing. Supported by fire from the infantrymen in foxholes and the squadron's mortars, this tactic usually disrupted the enemy attack. After suffering heavy casualties in several assaults against the 4th Cavalry, the Viet Cong and North Vietnamese in the area avoided night attacks on the unit's positions, preferring to attack fixed installations and less adept relief forces. Cavalry squadron commanders learned to move after dark to night locations, with squads of infantry riding the personnel carriers. The M113's changed direction several times during the move, making it virtually impossible for the enemy to determine their destination. Thus unable to locate the unit exactly, the enemy could not attack.

The tactics just described paid off for the South Vietnamese 4th Cavalry on 16 September 1967, when the 3d Squadron, stationed at Tam Ky, was alerted to move south to Quang Ngai. (By late afternoon the move was canceled, and the unit remained in its compound.) Subsequent information revealed that the Viet Cong had planned a night attack by six companies to seize the provincial headquarters and a major bridge on Route 1 south of Tam Ky and to occupy Tam Ky itself. Learning that the 3d Squadron had been ordered to Quang Ngai, the Viet Cong considered the moment opportune and scheduled their attack for that night.

The attack began at 0200 with mortar fire, which was followed by a massive infantry assault on the provincial headquarters and the bridge. The 3d Squadron at the armor compound, two kilometers distant, also began to receive enemy mortar fire. It was normal for the squadron to leave some troops at the compound when it was on an operation, and it was these troops that the mortar attack was intended to hit. Notified of the heavy ground assault and realizing the urgency of the situation, crew members, dressed in steel helmets, flack vests, underwear and slippers, took only twenty minutes to get their fifteen tracked vehicles out of the compound and on the road. The squadron, in radio contact with the provincial headquarters, was informed that the forces at the bridge were holding their own but that the headquarters was in imminent danger of being overrun.

The armor vehicles turned left on Route 1 and headed north toward the sound of battle. Reaching the entrance to the headquarters, they wheeled hard left, with APC's coming on line and mortar tracks setting up to the rear. The entire line assaulted, machine guns opening fire simultaneously and headlights turned on as the APC's swept forward. The Viet Cong attack disintegrated, and those who survived fled from the onrushing vehicles. The cavalrymen scoured the battlefield for prisoners and enemy equipment. The night counterattack by the cavalry had been devastating; over two hundred of the enemy were killed; three members of the squadron were slightly wounded.

The 4th Armored Cavalry, one of the original South Vietnamese armored units formed in the early fifties, was by 1967 well trained and aggressively led. In addition, it operated in an area where U.S. forces had not been introduced. However, in other parts of Vietnam, with the buildup of U.S. and free world forces reaching 545,000 by December 1967, Vietnamese units were assigned the mission of supporting revolutionary development—the 1967 term for pacification. To carry out this difficult task the South Vietnamese forces were usually stationed in or near populated areas. Since U.S. forces were in the field seeking the large Viet Cong and North Vietnamese units, South Vietnamese troops, including armor, assumed primarily a defensive role; reaction to enemy initiatives was their only source of action. The result was predictable: Vietnamese armor became static and its contacts with the enemy infrequent. Armored forces lost the aggressiveness they had begun to cultivate, and advisers found it hard to get armor troops into action. The battles of 1968 would eventually change this situation.

Improvements in Equipment

Paralleling organizational changes in the armored force of the South Vietnamese Army were important changes in materiel. Improvements were sometimes made in the United States and the equipment was sent to Vietnam for testing; more often, however, the ideas originated and the work was done in Vietnam to meet a specific need. The M113 was a success but some other armored vehicles were tried out and eventually discarded. The best example of the latter was the M114 armored reconnaissance vehicle, which was introduced into the U.S. Army in 1961 and 1962.

A U.S. Army team conducted staff briefings, driver training classes, maintenance courses, and demonstrations of the M114 for South Vietnamese soldiers. Graduates of these courses then organized and conducted classes in their own units and at the South Vietnamese Armor School. A total of eighty M114 reconnaissance vehicles were used to equip four newly organized armored cavalry reconnaissance squadrons. Others were issued to the Armor School and technical service installations for instructional and logistical purposes.

For a year the Army Concept Team in Vietnam evaluated the combat actions of the armored cavalry reconnaissance squadrons equipped with the M114. Although the squadrons had a few organizational and logistical problems, their critical problem was that the M114 could not move cross-country and had difficulty entering and leaving waterways. Since these defects limited the usefulness of M114's in Vietnam, the armored reconnaissance vehicles were replaced by M113's. Most Vietnamese reconnaissance squadrons began transition training with the M113 in April and finished by November 1964. Unfortunately for the service, the M114's lackluster Vietnam performance was ignored by U.S. Army decision makers and the vehicle with all its inadequacies became standard issue for the Army everywhere but in Vietnam. It was not until 1973 that General Abrams, then the U.S. Army Chief of Staff, branded the vehicle a failure and ordered it retired from the Army.

The M113, on the other hand, was the best land vehicle developed by the United States. With the aid of the M113, armored units could abandon the French tactic of sticking to the roads. An influx of enthusiastic U.S. advisers provided the spark needed for innovation. No armored vehicle is invulnerable, but the M113 proved to be as tough and reliable as any. It could absorb hits and continue operating. Only about one out of seven APC's penetrated by enemy fire was destroyed, and crew casualties were few. Successful as it was, however, the M113 did have limitations: to cut down on these and to provide new, more effective ways to use the vehicle, a number of modifications were tested.

The most necessary early modification was the provision of a gunshield for the APC's .50-caliber machine gun. The deaths of at least fourteen machine gunners at Ap Bac in early 1963 provided the incentive for the 2d Armored Cavalry, which fabricated the first two-gunners' shields, made of soft steel plating from the hull of a sunken ship. Later, when it was discovered that these could be penetrated, the Armor Command, the Advanced Research Projects Agency, and the Army Concept Team collaborated to make

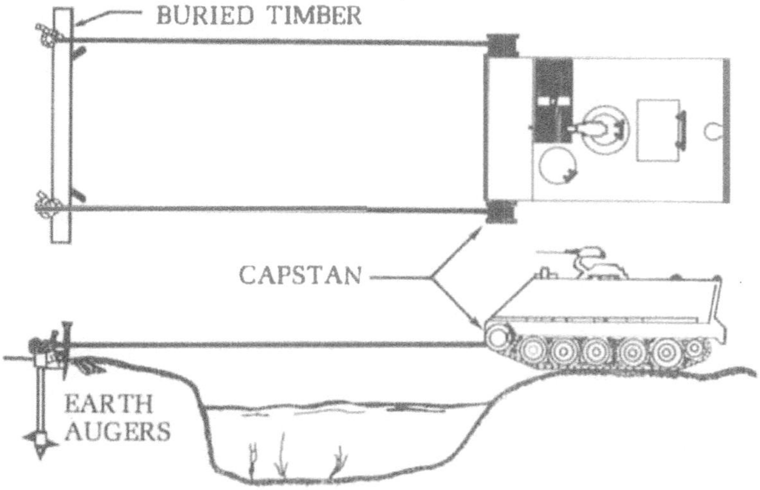

Diagram 1. Capstan and anchor recovery.

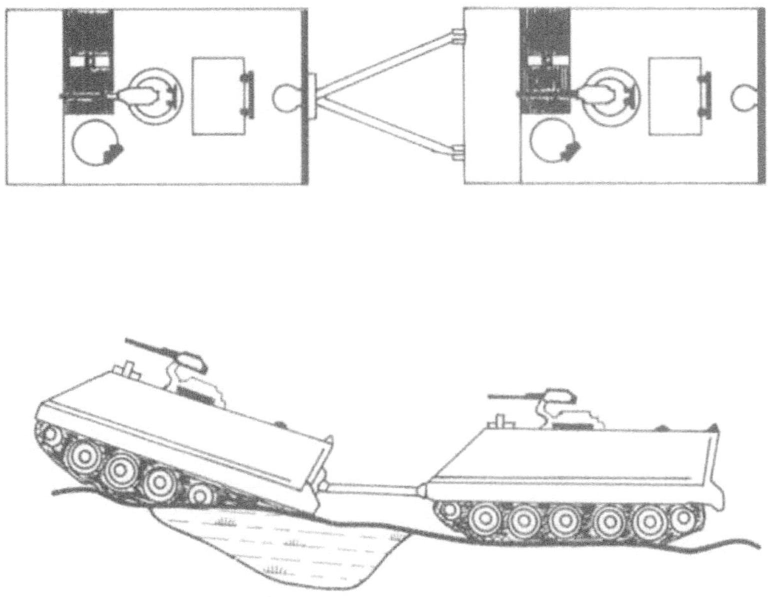

Diagram 2. Push-bar extraction. M113 employs a prefabricated bar to assist mired vehicles or improvises a push-bar from 4 x 4-inch timbers.

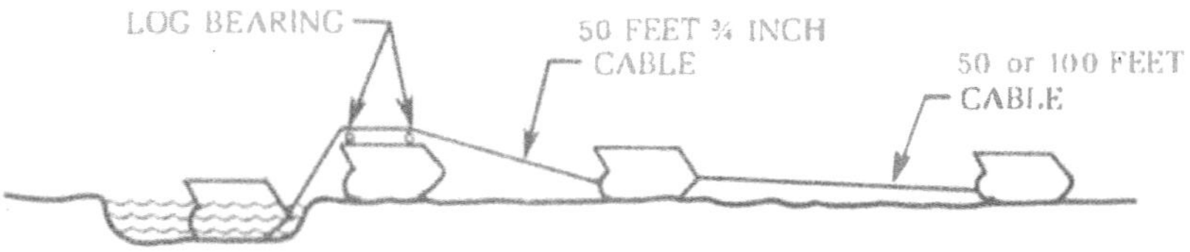

Diagram 3. Cable and log extraction is used to recover a vehicle from a canal with steep banks. The cable over vehicle 1 exerts an upward pull on the towed vehicle. Vehicles 2 and 3 exert forward pull. When the bow of the towed vehicle has been raised sufficiently, vehicle 1 moves forward. The logs on vehicle 1 serve as bearings.

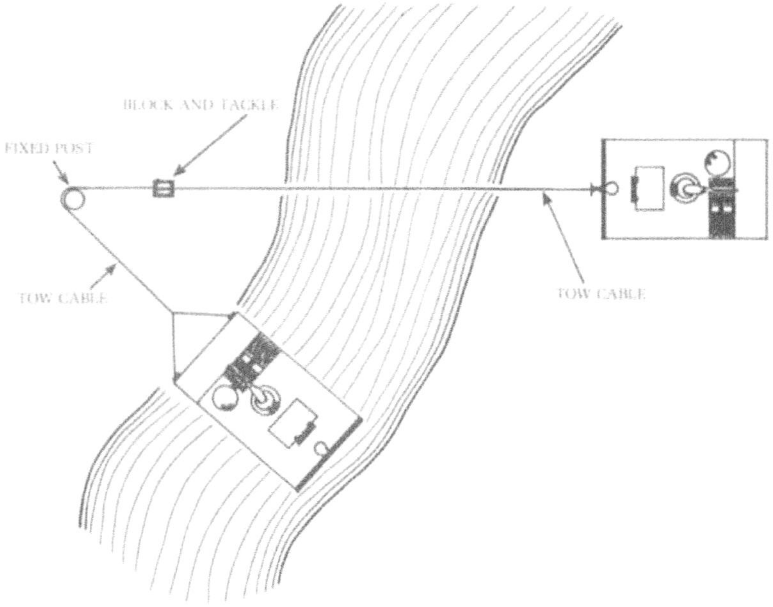

Diagram 4. Block and tackle.

gunshields from the armorplate of salvaged vehicles. These worked well, and with the assistance of the 80th Ordnance Depot in Saigon forty-six shields were distributed to mechanized rifle squadrons during the summer of 1963. After 1964 all APC's for the Vietnamese forces were fitted with these shields before being issued. The French had used similar gunshields on armored vehicles in Vietnam ten years before. Later, many armored personnel carriers were

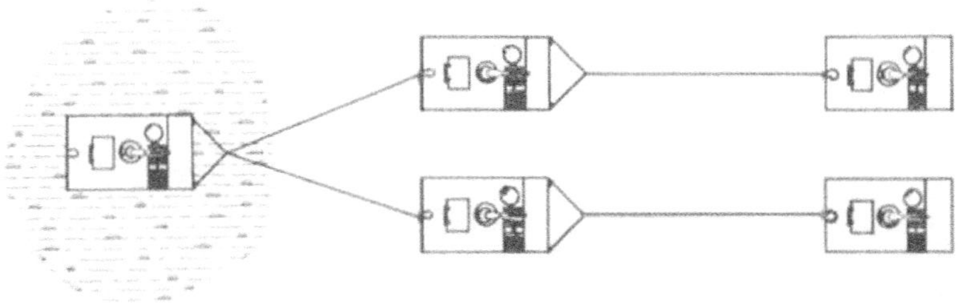

Diagram 5. Tow cables, a variation of the "daisy chain," used to extract an M113 mired in mud are 50 to 100 feet long. The yoke connections at the front of each vehicle are made from 10-foot cables issued with each M113. Poor traction near mired vehicle may require multiple tows as illustrated.

fitted with turrets mounting twin .30-caliber machine guns in place of the .50-caliber machine gun. Even 20-mm. guns were tried but were eventually discarded because of their mechanical unreliability.

Although the APC could travel across country, it did have trouble crossing the numerous canals, streams, and rivers in the delta. Of the several modifications devised, the capstan and anchor method of self-recovery proved to be one of the most successful techniques for getting an APC out of a steep-banked canal. Mechanized rifle squadrons operating in the delta devised several field expedients for canal crossing and vehicle recovery, among them the use of the push-bar, demolition, brush fill, 100-foot cable, and block and tackle. Even more rudimentary was the method known as the daisy-chain. A series of APC's were hitched together with cables and all the vehicles then pulled together. Muddy rice paddies were often crossed with as many as fifteen APC's chained together in parallel columns.

The average time a squadron took to cross a stream, however, was still considered excessive. Regimental commanders and advisers wanted to be able to begin crossing in five minutes or less after arrival. Although it did not meet the five-minute requirement, an aluminum balk bridge that could be placed over a thirty-foot gap by an M113 was developed in the latter half of 1965. With two additional balks, the bridge could be used by the M41 tank. Production of twenty-four M113 balk bridges was undertaken by the South Vietnamese Army engineers.

Originally, the M113, or APC, was intended merely as a troop carrier—a means of transport. Doctrine stated that the infantry were to dismount and engage the enemy. It soon became apparent

BALK BRIDGE CARRIED BY M113 *is demonstrated by troops of South Vietnamese Armor School.*

that the firepower of the .50-caliber machine gun, coupled with the vehicle's armor protection and mobility, produced a shock effect on the enemy. Some advisers and commanders realized that since the Viet Cong had no effective weapons to fight armor the M113 could be used as a mounted armored fighting vehicle. Consequently, Vietnamese troop commanders and U.S. advisers expressed a need for still more armament. At first, more machine guns were mounted on the sides of the vehicle by tying ground-mount tripods to the tops of APC's, then various fabricated mounts were tested and evaluated. In late 1963 it was common to see personnel carriers armed with side-mounted machine guns equipped with shields. In late 1964, even 2.75-inch rocket pods were attached to the sides of the carriers, but this scheme was abandoned because the fire was inaccurate. By 1965 the majority of the South Vietnamese cavalry squadrons used their vehicles in this manner, rarely dismounting, and then only when forced to do so by the enemy or the terrain. The APC's thus functioned as armored fighting vehicles, more like tanks than troop carriers.

When the enemy began to use weapons that were more effective against armor, the Americans and the South Vietnamese attempted to reduce the M113's vulnerability to penetration. Ammunition boxes, sand bags, and track blocks were hung on sides of vehicles. In 1966 closely spaced steel bars were installed on M113's of the 10th Armored Cavalry Regiment, and the Army Concept Team was slated to evaluate their use. Since no significant engagement took place during the observation period, however, no conclusions could be drawn as to whether the bar armor offered protection against antitank rockets and grenades. It was, however, easily damaged when struck by brush and other obstacles. It also slowed the vehicle down, and made it less maneuverable. The Army Concept Team decided the bar armor was inadequate and the program to install it was terminated.

To improve the troop protection of the M113, firing observation port kits were designed for M113's, and six modified vehicles were assigned to the 2d Armored Cavalry Regiment at Vinh Long in IV Corps Tactical Zone. Each M113 had a total of ten firing ports, four on each side and two in the rear. Above each firing port was an observation vision block. During the two-month evaluation made by the Army Concept Team no significant materiel deficiencies were noted. Because of Vietnam's consistently hot days, the absence of an enemy air threat, and the possibility of mines, South Vietnamese troops preferred to ride on top of the vehicle rather than inside. When the enemy was sighted troops jumped inside and fought, using chiefly machine guns atop the M113. After the test was completed in October 1966 the evaluators concluded that the modifications did not handicap the M113, but since the Vietnamese usually fought from the top with the hatches open there was no need for firing ports.

The test results were obviously biased by the fact that the team that evaluated the vehicles operated mainly with troops who fought on the top of the M113's. Controversy over the proper positioning of M113 crews and squads continued throughout the Vietnam War. Several units, mostly American, required crews to ride inside their M113's; others, pointing out the danger from mine explosions, did not. The differences of opinion remained unreconciled.

With the early success of the M113 mechanized rifle squadrons, it was only a matter of time until the M113 was modified for other roles. As early as August 1962, a flamethrower was installed in an M113, but it was used only four times in combat in a year. In December 1964, two M132 flamethrower vehicles were sent to the 1st Armored Cavalry where they were used aggressively in daily operations. After an evaluation the Army Concept Team recommended that four M132's plus two M113 service units be issued to each regiment.

The 81-mm. ground-mounted mortar was made organic to the

M41 IN SOUTH VIETNAMESE TRAINING OPERATION. *Tank has 76-mm. gun, light armor, two machine guns, and crew of four.*

armored cavalry regiment during the reorganization of November and December 1962. For the first few months of 1963 the use of the mortar was limited because the enemy fled when faced by M113's. As the Viet Cong learned how to deal with armor and began to engage the armored personnel carriers, mortar sections had more opportunities to support the troops. The need for a mounted mortar was soon apparent as units found it difficult to fire mortars from soft ground during the rainy season. Setting up ground mortars was also too time-consuming.

After evaluating the 81-mm. mortar section of the mechanized rifle squadron, the Army Concept Team recommended that an armored carrier such as the M106 with a 4.2-inch mortar be substituted. This recommendation was based on the need for the greater range and bursting radius that the 4.2-inch shell afforded. In May 1965, four M106 mortar carriers were permanently assigned to the 2d Armored Cavalry Regiment as a mortar troop. Evaluating this organization, the team concluded that although certain improvements were needed the weapon would provide excellent fire support and that one troop should be assigned to each regiment.

Tanks were also a problem, for by 1964 most of the M24 tanks left by the French had deteriorated into maintenance headaches. Repair parts, no longer in the U.S. supply system, were difficult to obtain. Engines, for example, were sent to Japan for rebuilding, but even this repair was uneconomical. The mechanical problems, coupled with the M24's inability to move cross-country, helped make South Vietnamese tank squadrons ineffective.

In mid-1964 the M41A3 tank was chosen to replace the M24, and the first M41's arrived in January 1965. Instructor training by a mobile training team from the U.S. 25th Division in Hawaii was completed on 17 April. Five squadrons were equipped and trained by the end of 1965. Although the first plan was to turn in the old M24's, the relics became pillboxes at installations throughout South Vietnam, except for a few under control of the Vietnamese Air Force at Tan Son Nhut.

The singularly outstanding and most consistently praised characteristic of South Vietnamese armored troops was their ability to perform individual and unit maintenance on vehicles and weapons; advisers commended them for keeping equipment operational with very limited support. Without recovery vehicles, armored units became extremely inventive. Since the supply system in the South Vietnamese Army was universally poor, both advisers and troopers became adept at "scrounging" replacement parts. Squadron and regimental mechanics performed such tasks as internal repair of starters, generators, radiators, and carburetors—maintenance normally accomplished by ordnance units. Deprived of aluminum welding, troops repaired holes and cracks in the hulls of APC's with wooden pegs and cement. Banana stalks and ponchos were used to mend radiators in water-cooled vehicles, and roadwheels were changed by putting the damaged wheel over a hole to relieve the ground pressure.

With an inadequate supply system, no turret or support unit mechanics, and the only replacement vehicles in Saigon, units still managed to field consistently over 90 percent of their equipment. Combat unit maintenance remained at a high level throughout the conflict.

Enemy Reaction to Armored Vehicles

Perhaps the best way to judge the success of the M113 is to examine the enemy reaction to it. The Viet Cong were not prepared for M113's when the South Vietnamese first used them in mid-1962, as the following passage from a captured document indicates: "The enemy APC's appeared while we were weak and our antitank weapons were still rare and rudimentary. We had no experience in attacking the APC. Therefore, the enemy's M113 APC's were effective and caused us many difficulties at first." When confronted by armored vehicles, the Viet Cong usually fled rather than fight; at first there were few attempts, if any, to engage mech-

M113 After Hits by Viet Cong 57-mm. Recoilless Rifle

anized units in open combat. Recognizing that they would have to adjust their tactics, the Viet Cong began to train to attack mechanized vehicles. The Viet Cong doctrine stressed occupation of dug-in positions in the face of APC assaults, and enemy soldiers were soon learning crude methods of destroying the M113. Holes the size of an APC, nicknamed tiger traps by advisers, were soon found in delta roads. Observing the difficulties that APC units had in crossing canals, the Viet Cong used canals as obstacles in their positions, and frequently they mined possible crossing sites.

In 1965 the Viet Cong published a comprehensive and fairly accurate training document entitled Attack on M113 APC. This document listed characteristics of the APC; organization, equipment, and strength of mechanized units; tactics used by APC units; methods of attacking an APC; and some training techniques. Included were instructions for using the new antitank weapons. In the spring of 1963 the Viet Cong had begun to use recoilless rifles with 57-mm. high explosive, antitank rounds, and the number of hits on M113's had increased dramatically by the fall of 1963. Although the rounds often penetrated, they did not usually destroy the M113. Late in that year, armor-piercing .30-caliber ammunition along with a large number of automatic weapons was found in a Viet Cong cache in the delta. The extent of the Viet Cong antiarmor equipment became apparent when, in August 1963, an armor-piercing grenade was discovered. The first 75-mm. recoilless rifle was captured in September, and in December the first M113 was damaged by a 75-mm. round. This arms buildup continued into 1964, when a variety of mines, both pressure and electrically detonated, accounted for the majority of damaged and destroyed vehicles.

By 1965 the Viet Cong was using armor-defeating weapons as low as company level among regular and provincial units. Newly organized weapons platoons, companies, and battalions armed with 57-mm. and 75-mm. recoilless rifles and .50-caliber machine guns were issued the rocket propelled antitank grenade, RPG2. For several years this weapon with its B40 warhead was the principal enemy weapon against armor. Eventually, the RPG2 was replaced by the RPG7, an improved antitank grenade with a more lethal warhead, greater range, and a better sight. This weapon and the antivehicular mine were the enemy's most successful antiarmor devices and were a constant and persistent problem for South Vietnamese and allied armor throughout the conflict. Although many measures to defeat these weapons were tried, no adequate means was ever found.

"Coup Troops"

Any discussion of the armored forces of the Republic of Vietnam in the early 1960's would be incomplete without some reference to the effect of politics on armored units. Political-military relationships in Vietnam had always been highly complex, but during 1963 the South Vietnamese armored units emerged as an important factor in the political-military equation. Since armor had been used to suppress at least three attempted *coups d'etat,* it was believed that armored forces had the power to uphold or depose any government. The 1st Armored Cavalry Regiment in Saigon and the 2d Armored Cavalry in My Tho, about sixty kilometers to the south, could be combined on short notice to support or suppress an attempt to overthrow the government.

President Ngo Dinh Diem took steps early to insure the loyalty of his armor leaders. Only the most trustworthy armor officers were assigned to the greater Saigon area. Diem also severely limited fuel allocations, so that most armored units could not reach Saigon without additional fuel. Under this system extended armor operations against the enemy were impossible. Before the November 1963 coup, President Diem and his brother Nhu developed elaborate means of monitoring troop movements. For example, Captain Ly Tong Ba, commander of an armored unit in My Tho, had to telephone the palace frequently to report his location.

In spite of President Diem's vigilance, on 1 November 1963 armored units from all over South Vietnam, including units from the Armor School, converged on the presidential palace to reinforce the demand that Diem resign. Early the next morning, tanks and infantry assaulted the palace, which was defended by tanks of the presidential guard brigade. The guards resisted valiantly until, hopelessly outnumbered and with several of their tanks in flames, they surrendered.[4] Diem, meanwhile, escaped but was captured in the Cholon section of Saigon. While en route to Army headquarters, he and his brother were assassinated in an M113 of the 4th Squadron, 4th Cavalry.

Many armored officers were praised for their participation in the coup and rewarded with promotions. The tanks, APC's, and armored cars that participated in the coup proved the power of armored units as supporters of a political group. After the Diem downfall, the political situation was unstable, and subsequent governments kept a close account of key units that could support or

[4] Oddly enough, the commander of the palace armored unit was promoted from captain to major on the spot, an unusual reward for loyalty to a losing cause in a revolt.

defeat a new coup. As early as 30 January 1964, tanks participated in General Khanh's seizure of power.

In September, after an abortive coup attempt, there was a command reorganization that resulted in a number of officer reassignments and the imposition of restrictions on the movement of armored vehicles and units. After several South Vietnamese Armor School officers were accused of involvement in another abortive coup on 19 February 1965 the school was ordered to move. Leaving the advisers at Thu Duc, the school departed for Van Kiep on 25 February. The Vietnamese high command soon reversed its decision, and two weeks later the school returned to Thu Duc. For some time thereafter the number of forces located at the Armor School was closely scrutinized.[5]

Political appointees as commanders were a fact of life until the mid-1960's and had a disastrous effect on the development and use of armor. These commanders were not always the best leaders, and often were completely lacking in initiative. Since many of these men were concerned only with their own careers, the training and combat operations of armored units suffered accordingly. If a unit had too many casualties and lost too many vehicles, particularly during Diem's rule, the commander was likely to be relieved. As a result, commanders, even the good ones, were reluctant to fight the enemy for fear of losing their jobs. Because of the necessity to protect political regimes, armored units, particularly tank squadrons, were seldom allowed far from the ruling headquarters. Thus, even down to the province level, the term palace guard was applied to many armored units.[6] While the coups were being planned, armored units were not fighting the enemy. Only when political stability became a reality in 1968 were armored units finally able to shed this stigma of palace guard.

---

[5] The Armor School was forbidden even to conduct training in the direction of Saigon. One night, when U.S. advisers were bringing in new M-41 tanks after midnight to avoid traffic, General Khanh was alarmed and fled with his family to Vung Tau, over 50 kilometers away.

[6] Armored units were often given the name "coup troops" by detractors. In the same vein, the tanks were called "voting machines" because they influenced every early change of government. Although the armor corps became nonpolitical, the U.S. Embassy still worried that during the 1971 Vietnamese national elections armored forces would intervene.

CHAPTER III

Growth of U.S. Armored Forces in Vietnam

The American elections of 1960 brought John F. Kennedy to the White House and Robert S. McNamara to the Pentagon. The change spelled the end of the strategy of massive retaliation and of the pentomic division with its five battle groups designed to fight nuclear wars. The Army reorganization of 1963 restored the infantry battalion and provided a structure for the whole Army that, at battalion and brigade level, was much like the separate battalions within the combat commands of U.S. armored divisions after World War II. It was clear that the new policy of flexible response demanded a force that could fight in any kind of war, including so-called wars of national liberation.

In armored units there was little change. The 1963 reorganization reduced each tank and mechanized infantry battalion to three line companies, but each division had more battalions and support echelons. No one in armor seriously believed that armored unit tactics needed to change. In 1957 Field Manual 17-1, Armor Operations, Small Units, devoted only two and one-half pages to guerrilla warfare. By the early 1960's that coverage had been broadened; Field Manual 17-35, Armored Cavalry Platoon, Troop and Squadron, carried an expanded treatment of guerrilla fighting under the title, Rear Area Security.

Many of the tactics set forth in the manual for employing armored cavalry in rear area security missions proved useful in Vietnam. Road security, base defense, air reconnaissance, reaction forces, and convoy escort were described. Field Manual 17-1 included discussions of base camps, airmobile forces, tailoring of forces for specific missions, encirclements, and ambushes. Both books stressed surveillance, the use of the combined arms team, and the need for mobility. Yet most counterinsurgency training was limited to work on patrols, listening posts, and convoy security; the Army did not foresee a whole theater of operations without a front line or a secure rear area.

Although the helicopter was not specifically designed for counterinsurgency warfare, it proved to be one of the most useful machines the U.S. Army brought to Vietnam. As early as 1954 the Army had studied the use of helicopters in cavalry units, and later experiments with armed helicopters had been conducted at the

U.S. Army Aviation School at Fort Rucker, Alabama. By early 1959 the U.S. Armor School at Fort Knox, Kentucky, and the U.S. Army Aviation School had developed an experimental Aerial Reconnaissance and Security Troop—the first air cavalry unit. This aerial combined arms team, composed of scouts, weapons, and infantry, was tested in 1960 and recommended for inclusion as an organic troop in divisional cavalry squadrons. In early 1962 the Army's first air cavalry troop, Troop D, 4th Squadron, 12th Cavalry, was organized at Fort Carson, Colorado, with Captain Ralph Powell as its commander. The troop mission was to extend the capabilities of the squadron in reconnaissance, security, and surveillance by means of aircraft. Over the next three years all divisional cavalry squadrons in the Army were provided with air cavalry troops.

In mid-1962 Lieutenant General Hamilton H. Howze headed a study group to examine the possibilities of the helicopter in land warfare. The group concluded that helicopters organic to the ground forces were an inevitable step in land warfare. The Howze Board foresaw air assaults, air cavalry operations, aerial artillery support, and aerial supply lines, and recommended the creation of an air assault division. The outcome of the study was the formation of the 11th Air Assault Division, later to become the 1st Cavalry Division (Airmobile). The division organization included one unique unit, an air cavalry squadron made up of one ground troop and three air cavalry troops.

By 1965 when the U.S. Army began to send units to Vietnam, divisional armored cavalry squadrons had three ground cavalry troops and an air cavalry troop, tank battalions had three identical tank companies, and mechanized infantry battalions had three mechanized companies mounted in APC's. *(Chart 1)* Armored units were equipped with a mixture of M48 and M60 tanks, M113 armored personnel carriers, and M109 self-propelled 155-mm. howitzers.

On the eve of the Army's major involvement in Vietnam, however, most armor soldiers considered the Vietnam War an infantry and Special Forces fight; they saw no place for armored units. The Armor Officer Advanced Course of 1964–1965 never formally discussed Vietnam, even when American troops were being sent there. Armor officers were preoccupied with traditional concepts of employment of armor on the fields of Europe; a few attempted to focus attention on the use of armor in Vietnam, but in the main they were ignored. Many senior armor officers who had spent years in Europe dismissed the Vietnam conflict as a short, uninteresting interlude best fought with dismounted infantry.

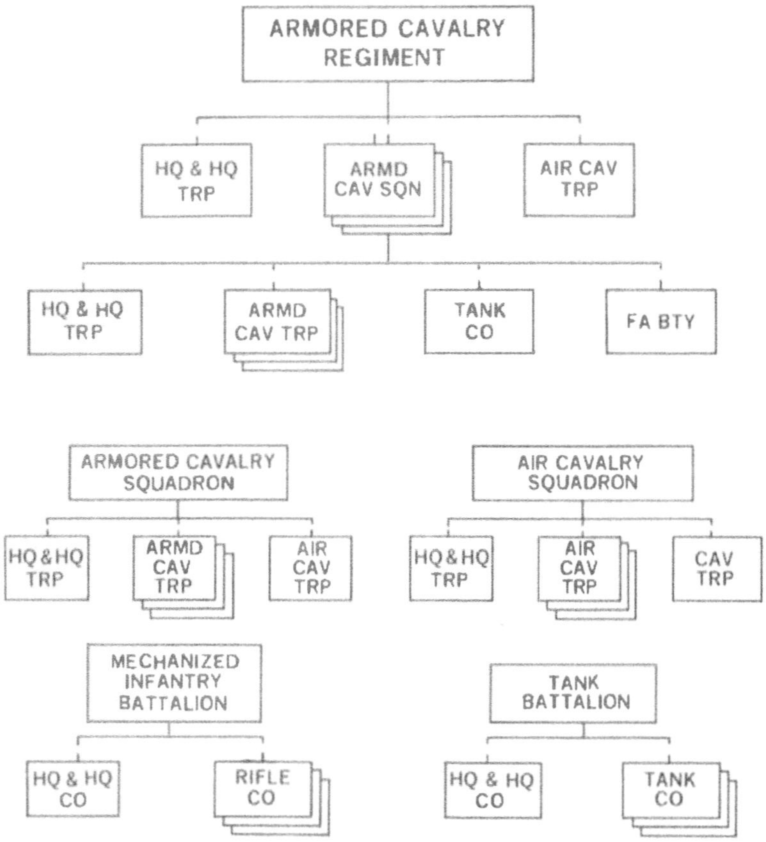

CHART. U.S. ARMORED ORGANIZATIONS, 1965

The Marines Land

By early 1965 the American command in Vietnam had concluded that the South Vietnamese could not hold off the combined Viet Cong and North Vietnamese Army forces without U.S. assistance, and in February American forces began a limited air and sea bombardment of North Vietnam and jet aircraft strikes in South Vietnam. In late February General William C. Westmoreland, Commander, U.S. Military Assistance Command, Vietnam, requested two U.S. Marine battalion landing teams to assist South Vietnamese forces in making the airfield at Da Nang secure.

On 9 March 1965 Marine Corps Staff Sergeant John Downey

U.S. MARINE CORPS FLAMETHROWER TANK IN ACTION NEAR DA NANG. *The U.S. Army no longer had flamethrower tanks but the Marine Corps sent several to Vietnam.*

drove his M48A3 tank off a landing craft onto Red Beach 2 at Da Nang and was followed by the rest of the 3d Platoon, Company B, 3d Marine Tank Battalion, the first U.S. armored unit in Vietnam. Later in March the 1st Platoon, Company A, 3d Marine Tank Battalion, landed with a second Marine team, and for the remainder of the month these two platoons bolstered the defenses of the Da Nang airfield.

On 8 July 1965 the 3d Marine Tank Battalion, commanded by Lieutenant Colonel States R. Jones, Jr., debarked at Da Nang—the first U.S. tank battalion in Vietnam. Consistent with U.S. Marine Corps concepts of tank employment, the battalion's primary mission was to support Marine infantry. As Marine tactical areas of responsibility expanded throughout 1965, so did the areas in which tank units operated in support of infantry reaction forces, as support for infantry strongpoints, or in sweep and clear operations.

U.S. Marine Corps tank units took part in the first major battle involving U.S. armored troops—Operation STARLITE. In mid-August

1965 a large Viet Cong force believed to include the 1st Viet Cong Regiment was reported to be preparing to attack the Chu Lai airfield, southeast of Da Nang. To preempt the attack, a marine amphibious operation using helicopters and three battalion teams was mounted. A tank platoon supported each battalion. The operation lasted for two days, and was characterized by short, bitter firefights as the Viet Cong attempted to evade encircling forces. When Operation STARLITE ended, the marines had pushed the Viet Cong regiment against the sea, killing over 700 men. In two days, seven tanks had been damaged by enemy fire, three so badly that the turrets could not traverse, and one to the point that it had to be destroyed by a demolition team. The Marine tank crews had demolished many enemy fortifications, captured twenty-nine weapons, and killed sixty-eight Viet Cong.

Until their final withdrawal in late 1969, Marine Corps tanks continued to support U.S. Marine Corps infantry units in Vietnam. During this time, Marine armor, eventually two full battalions, participated in the battles of Khe Sanh and Hue and was the main armor defense force of the Demilitarized Zone in the north.

Decision Making

When the decision to send American ground forces to Vietnam was finally made after a long, involved debate at high governmental levels it was conditional. Troops were released to Vietnam in increments, each designed to support one of the three principal ground strategies that followed one another in rapid succession. The cumulative effect of rapidly changing strategy and the absence of a clearly stated long-term goal with a definite troop commitment can be easily seen in retrospect, but in the hectic days and months of the first half of 1965 there was no one who could predict the length of the war, enemy intentions and capabilities, and the extent of future U.S. commitment. It was in this atmosphere that General Westmoreland and his planners labored to develop troop lists of units they wanted sent to Vietnam.

During the first half of 1965 the three principal ground strategies were described as security, enclave, and search and destroy. Of the first two, neither required the use of large mobile forces nor implied that U.S. troops would stay in Vietnam for any length of time. Under the security strategy American marines were sent to Vietnam to defend an airfield and took their tanks along. The planners in the U.S. Military Assistance Command, Vietnam, had not examined the makeup of a marine battalion landing team and therefore did not realize that the team had tanks. The marines, for their part, saw no reason to leave their tanks behind. That tanks were sent to Vietnam was, therefore, a kind of accident, and a not altogether popular one. For example, U.S. Ambassador Maxwell D. Taylor was surprised and displeased to learn that the marines had landed with tanks and other heavy equipment "not appropriate for counterinsurgency operations."

The security strategy was defensive in intent and limited in scope. In neither it nor its successor, the so-called enclave strategy, was the use of armored forces planned. The troops sent to carry out both strategies—the Marines, the U.S. Army 173d Airborne Brigade, and an Australian infantry battalion—were therefore dismounted infantry elements whose stay in Vietnam was considered temporary. In fact, the airborne troops were sent to Vietnam on temporary duty. Force planners, trying to get all the combat troops on the ground they could and still stay within the limited troop ceilings and the very restricted capacity of the logistical base, chose infantry units that were easily deployed and required only minimal support.

The third strategy, search and destroy, began to evolve during June and July of 1965. After receiving the authority to use U.S. ground forces anywhere in Vietnam in search and destroy operations, General Westmoreland tried to determine the forces that would be necessary to defeat the enemy. Throughout 1965 he personally studied every unit on the troop lists to insure the best use of the authorizations, the earliest possible troop deployments, and the most appropriate apportionment of troops among the armed services. Within Army manpower authorizations, he also sought an effective distribution among branches. The MACV staff continuously reevaluated the situation to determine whether there was a need for additional troops. Because of the troop ceilings—the first three troop increments were only 20,000 each, with no promise of more—the severely limited logistical base, and the many misconceptions about the country, armored units were not seriously considered for early employment in Vietnam.

The first debate on the use of armored units arose during planning for the deployment of the 1st Infantry Division. Directives from the Department of the Army required reorganization of the division to eliminate the units designed for nuclear war, the division's two tank battalions, and all the mechanized infantry. The mechanized infantry units were to be organized into dismounted infantry battalions. The rationale for these decisions was provided by the Chief of Staff of the Army, General Harold K. Johnson, in a message on 3 July 1965 to General Westmoreland. Overruling an Army staff proposal that one tank battalion be retained, General Johnson gave some of his reasons.

A. Korean experience demonstrated the ability of the oriental to employ relatively primitive but extremely effective box mines that defy detection. Effectiveness was especially good in areas where bottlenecks occurred on some routes. Our tanks had a limited usefulness, although there are good examples of extremely profitable use. On balance, in Vietnam the vulnerability to mines and the absence of major combat formations in prepared positions where the location is accessible lead me to the position that an infantry battalion will be more useful to you than a tank battalion, at this stage.

B. I have seen few reports on the use of the light tanks available to the Vietnamese and draw the inference that commanders are not crying for their attachment for specific operations.

C. Distances and planned areas of employment of the 1st Division are such that the rapid movement of troops could be slowed to the rate of movement of the tanks.

D. The presence of tank formations tends to create a psychological atmosphere of conventional combat, as well as recalls the image of French tactics in the same area in 1953 and 1954.

General Johnson went on to say that the divisional cavalry, the 1st Squadron, 4th Cavalry, would be allowed to keep its medium tanks, M48A3's, to test the effectiveness of armor. If circumstances required it later, General Johnson was prepared to reinforce the division with a tank battalion. In his answer to this message General Westmoreland declared that "except for a few coastal areas, most notably in the I Corps area, Vietnam is no place for either tank or mechanized infantry units."

These two messages clearly show the prevailing attitude among American senior officers toward the use of armored forces in Vietnam, and reflect the influence of the French experience with armor. At the staff level, the commanders' misconceptions were magnified by problems of force structure; troop ceilings and the limited logistical base became further justification for rejecting armored units. For example, when it was noted that a mechanized battalion required more than 900 troops and a dismounted infantry battalion only about 800, dismounted infantry became the choice. Further, the mechanized battalion also needed a direct maintenance support unit of over 150 troops and a security force to guard its base. Although a tank battalion required but 570 men, its detractors were quick to say that only 220 of these were fighters—the rest were support troops. To support the tank battalion in combat still more maintenance and security troops would be needed. This line of reasoning made the tank battalion an unattractive package. One force planner in the Military Assistance Command, Vietnam, commented that while no one was outspokenly prejudiced against the use of armored forces in Vietnam there was no strong voice calling for their employment.

The experience of the 1st Infantry Division illustrates some of the problems faced by the commanders of armor units earmarked for Vietnam. Having lost his tank and mechanized infantry units, Major General Jonathan O. Seaman, the 1st Division commander, wanted to make sure that his one remaining armored unit, the 1st Squadron, 4th Cavalry, was properly equipped. General Seaman, noting the poor performance of the M114 reconnaissance vehicle in Vietnam, recommended substituting M113's for the squadron's M114's. After considerable resistance from the Army staff in Washington the exchange was finally approved. The M113's, modified with additional machine guns and gunshields, came from the pool of vehicles taken from the recently dismounted mechanized battalions. Obtaining trained people was another matter. Some months before when General Seaman had been told that only one of his brigades would go to Vietnam, he had filled the deploying brigade with experienced officers and men from other division units, including the 4th Cavalry. Subsequently the remainder of the division was ordered to Vietnam. Filled with new officers and troops the cavalry had time for only two weeks of unit training before it left.

When the first armor units reached Vietnam their tactical employment was equally frustrating for the squadron and battalion commanders: again the experience of the 1st Squadron, 4th Cavalry, is illustrative. The brigades of the 1st Division, each with a cavalry troop attached, were sent to three different locations. The squadron headquarters was left with only the air cavalry troop under its control. The first operation at squadron level took place in early 1966, over six months after the unit reached Vietnam. Cavalry tactics during those first six months suffered from the "no tanks in the jungle" attitude. Because General Westmoreland saw no use for tanks, the M48A3's were withdrawn from the cavalry troops and held at the squadron base at Phu Loi. It took six months for General Seaman and Lieutenant Colonel Paul A. Fisher, the squadron commander, to convince General Westmoreland that tanks could be properly employed on combat operations. Although the tank crews subsequently proved themselves in combat, General Seaman's repeated requests for one of his tank battalions that had been left at Fort Riley were refused. The same fate befell similar requests by his successor throughout 1966.

The air cavalry troop was also fragmented. Before the unit was sent to Vietnam it was organized like an armed helicopter company: a troop of two platoons, equipped with command and control helicopters and gunships. Training was based on experience with armed helicopters in Vietnam. Once in Vietnam, however, helicopters were parceled out to the units that wanted them, and the aerorifle platoon was attached to Troop C for use as long-range patrols and base camp security guards. The troop continued to operate in this makeshift fashion for over a year until the example of other air cavalry units brought about a change in organization and tactics.

As the political and military power of the enemy continued to grow during 1965, General Westmoreland and his planners were convinced that the United States would have to provide additional troops if the government of South Vietnam was to survive. When the evolving strategy required additional American forces, the President of the United States increased manpower authorizations, and eventually more armored units were sent to Vietnam. In late 1965 General Westmoreland requested deployment of the 11th Armored Cavalry and the 25th Infantry Division. Included in the latter were the 3d Squadron, 4th Cavalry, the 1st Battalion, 5th Infantry (Mechanized), and the 1st Battalion, 69th Armor. Major General Frederick C. Weyand, the 25th Division commander, insisted on deploying the armor battalion despite resistance from staff planners in both Department of the Army and Vietnam.

Scouts Out

In the U.S. cavalry of the late 1800's, the familiar call at the beginning of a campaign was "scouts out"; so it was, too, in Vietnam, in 1965. Earlier some ground cavalry units had been used in Vietnam, but in September the first air cavalry unit, the 1st Squadron, 9th Cavalry (Air), arrived with the 1st Cavalry Division (Airmobile). While a brigade of the 101st Airborne Division maintained security in the An Khe camp area, the air cavalry troopers, together with the airmobile infantrymen of the division's traditional "cavalry" battalions, were allowed a few days to move into and secure an area for building a division base.[1] By 18 September enough aircraft were available for the squadron to begin aerial reconnaissance of the division base area.

In late October 1965 reports of increasing enemy strength in the central plateau brought commitment of the entire 1st Cavalry Division to an offensive in western Pleiku Province. The 1st Squad-

[1] Although battalions in the 1st Cavalry Division retained the traditional cavalry names, they were actually dismounted infantry units. Since the division was airmobile, these units moved by helicopter. The only unit in the division that fought mounted was the 1st Squadron, 9th Cavalry, an air cavalry squadron with one ground cavalry troop.

ron, 9th Cavalry (Air), the "eyes and ears" of the division, was given the mission of finding the enemy as well as covering divisional troop movements. Few enemy troops were spotted from the air at first but by 30 October the number of sightings began to increase. Troop C captured three North Vietnamese soldiers, the first divisional unit to capture North Vietnamese Army troops. On 1 November 1965 Troop B's scouts sighted and fired on eight North Vietnamese soldiers. Shortly thereafter about forty more were spotted and the troop's aerorifle platoon, already airborne, was landed to investigate. For once the enemy stood and fought: the American platoon killed five of the enemy and captured nineteen others. The troopers found the reason for the enemy concentration when they moved into a nearby stream bed and discovered a hospital. Fighting around the hospital continued while the captured soldiers, medical equipment, and supplies, all part of the 33d North Vietnamese Army Regimental Hospital, were evacuated. Late in the afternoon the 2d Battalion, 12th Cavalry (Airmobile) landed as reinforcements, and the 9th Cavalry aerorifle platoon was withdrawn.

Knowing that the enemy was in the area in strength, the 1st Squadron, 9th Cavalry (Air), with Company A, 1st Battalion, 8th Cavalry (Airmobile), moved to the Duc Co Special Forces Camp and by evening on 3 November 1965 had begun a reconnaissance in force along the Cambodian border. The squadron ambush force, consisting of three American aerorifle platoons, an attached Vietnamese platoon, and a mortar section of Company A, 1st Battalion, 8th Cavalry, reconnoitered and established three ambush sites. In the early evening the southernmost ambush, manned by Troop C's aerorifle platoon, sighted a large, heavily laden North Vietnamese Army company. The enemy soldiers, easily seen by the light of the full moon, were laughing and talking and obviously felt secure in that part of the jungle. The waiting cavalrymen detonated eight claymore mines set along a 100-meter kill zone, and the troopers joined in with their M16 rifles as additional claymores and rifle fire from the flank security elements sealed off the area. The firing lasted two minutes, and when there was no answering fire from the enemy the aerorifle platoon returned to the patrol base.

One ambush platoon was still in position at 2200 when the three platoons in the base were assaulted by an estimated enemy battalion. The first attack was costly to the North Vietnamese, who withdrew. Enemy snipers remained in the nearby trees, and, aided by bright moonlight, attempted to pick off the defenders. With the help of the reinforcements from Company A, 1st Battalion, 8th

Cavalry, the Americans defeated another enemy attack just before daylight. The defenders then made limited counterattacks to expand the perimeter and provide a safer landing for the rest of the battalion. By dawn the enemy began to break away, incoming fire slackened, and there was only occasional sniper fire from the surrounding trees. The air cavalry platoons were then extracted, leaving the 8th Cavalry to sweep the battlefield.

For the second time in a week air cavalry soldiers had successfully battled the enemy, later identified as the 66th North Vietnamese Army Regiment, recently arrived in South Vietnam. These combat actions and the scouting activities of the entire squadron supplied information on enemy locations that brought major elements of the 1st Cavalry Division into action for almost a month in the Ia Drang valley. The first stage of the Ia Drang campaign, which was also the first major battle for an air cavalry squadron, showed how the air cavalry should be used. For those commanders who employed it properly, the air cavalry in Vietnam was a primary source for gathering intelligence information.

Ap Bau Bang

As the 1st Squadron, 9th Cavalry, was triggering the Ia Drang campaign, far to the south Troop A, 1st Squadron, 4th Cavalry, participated in the first major engagement of the 1st Infantry Division. The battle at Ap Ban Bang was an early example of a combined arms defense of a night position. The action was important because it occurred during the initial stages of U.S. troop involvement and demonstrated the effectiveness of combined arms in jungle warfare.

A task force of the 2d Battalion, 2d Infantry, consisting of the battalion's three rifle companies, its reconnaissance platoon, and Troop A (less the nine tanks still at Phu Loi), was ordered to sweep and secure Highway 13 from the fire base at Lai Khe to Bau Long Pond, fifteen kilometers north. The purpose of Operation ROADRUNNER was to secure safe passage for a South Vietnamese infantry regiment and provide security for Battery C, 2d Battalion, 33d Artillery, which was moving north to support the South Vietnamese regiment. On 10 and 11 November 1965 the road was cleared without incident: medical teams even visited the village of Bau Bang as part of a medical civic action program.

During the afternoon of 11 November, Troop A, the artillery battery, the command group, and Company A of the infantry battalion moved into a defensive position south of Bau Bang. *(Map 5)* Concertina wire was installed, individual foxholes were dug, and

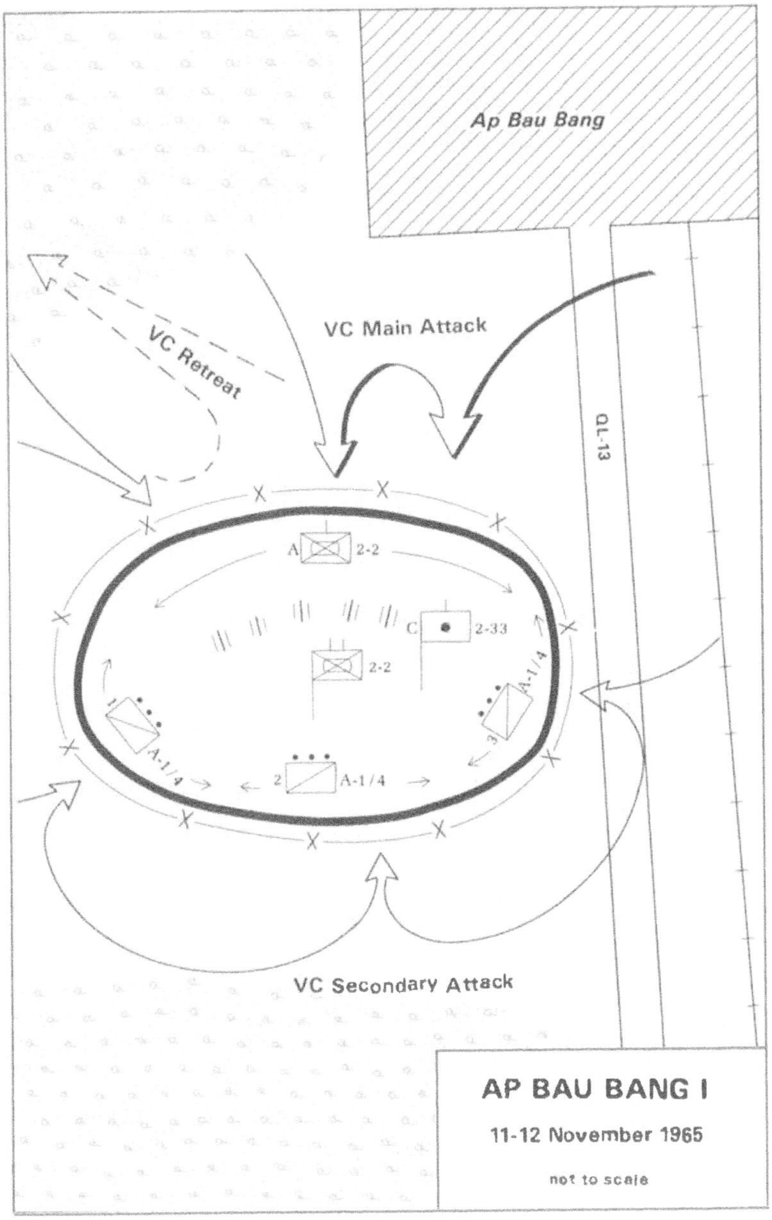
Ap Bau Bang
VC Retreat
VC Main Attack
QL-13
A 2-2
C 2-33
2-2
A-1/4
1
2 A-1/4
3
VC Secondary Attack
AP BAU BANG I
11-12 November 1965
not to scale

MAP 5

patrols were set up for ambushes. Dragging the hull of a destroyed armored personnel carrier around the perimeter, Troop A knocked down bushes and young rubber trees to clear fields of fire. The night passed with only a light enemy probe, but within minutes after the early dawn *stand-to* (a term applied by armored units to first-light readiness of men, vehicles, and radios) fifty to sixty mortar rounds exploded inside the perimeter. In the first few minutes Troop A had two men wounded. Half an hour later a violent hail of automatic weapons and small arms fire was added to the mortar fire. Under cover of this fire, the Viet Cong moved to within forty meters of the defensive positions. While the cavalrymen returned the fire, M113's of the 3d Platoon roared out and assaulted the enemy. The violence of this unexpected mounted counterattack disrupted the Viet Cong attack, and the M113's returned to the perimeter. The troop suffered three more wounded and one killed when ammunition in a mortar carrier exploded after being hit by enemy mortar fire.

The Viet Cong made their second assault from the jungle and rubber trees south of the perimeter. Again supported by mortars and automatic weapons, they crawled through the waist-high bushes of a peanut field and rushed the concertina wire. One of the M113's in that section of the perimeter was driven by Specialist 4 William D. Burnett, a mechanic. When the .50-caliber machine gun on his APC failed to function, Specialist Burnett jumped from the cover of the driver's compartment to the top of the vehicle, cleared the weapon, and opened fire on the charging Viet Cong, killing fourteen. For this and other actions during the battle, Specialist Burnett was awarded the Distinguished Service Cross. The heavy fire from Burnett's machine gun and those of the M113's near him broke the enemy assault.

The Viet Cong next attacked west from Highway 13, and again were repulsed by .50-caliber and small arms fire. Several times M113's were moved to weak points on the perimeter so that their machine guns could fire into the enemy's ranks at point-blank range. At 0645 an air strike directed by an airborne forward air controller dropped bombs and raked the wooded area north of the task force with 20-mm. cannon fire.

At 0700 the Viet Cong began their main charge from the north out of Bau Bang, supported by recoilless rifles and automatic weapons emplaced along an east-west berm on the southern edge of the village and mortars in the village itself. The main attack was stopped at the wire by the combined firepower of Company A and Battery C, which in thirty minutes, using two-second delay fuzes, fired fifty-five rounds point-blank at the attackers. Despite this wall of steel, one Viet Cong squad penetrated the perimeter and threw a grenade into a howitzer position, killing two artillerymen and wounding four others. Three infantry companies were meanwhile ordered by the 1st Infantry Division to move toward the battle and to envelop the Viet Cong from the rear. At the same time, armed helicopters flew to the scene.

The enemy attacked again at 0900, this time from the northwest, with recoilless rifles, automatic weapons, and mortars. Protected by the berm, these weapons could not be destroyed by direct artillery fire. When an Air Force pilot reported that the villagers were fleeing to the north of Bau Bang, and another pilot sighted the mortar positions within Bau Bang, permission was obtained to hit the village. Fighter planes bombed the enemy positions and armed helicopters discharged rockets and strafed. For the next three hours, while mortars, artillery, and air strikes hammered the enemy, the task force repelled successive attacks.

The battle of Ap Bau Bang went on for more than six hours before the enemy withdrew to the northwest, leaving behind his wounded and dead. Troop A, commanded by Second Lieutenant John Garcia, suffered seven killed and thirty-five wounded; two M113's and three M106 mortar carriers were destroyed and three M113's were damaged. Procedures and techniques learned in training had been proven in battle. The clearing of fields of fire and the pre-dawn stand-to had insured the full application of Troop A's fire-power. The 3d Platoon's foray into the enemy position and the positioning of M113's on the perimeter had demonstrated the unit's flexibility, and artillery and aerial fire support had provided depth to the defense. The enemy had begun the fight; the combined arms team had ended it.

Deployments and Employments

The 1st Squadron, 4th Cavalry, had made the case for armored forces. Upon the recommendations of General Seaman and others, armored units of the 25th Infantry Division were sent to Vietnam in early 1966. Before leaving, the 1st Battalion, 5th Infantry (Mechanized), had equipped its APC's with gunshields and extra machine guns and the 1st Battalion, 69th Armor, had exchanged its gasoline-powered M48A2C tanks for diesel-powered M48A3's. For these units and others to follow, South Vietnamese Armor Command advisers had prepared a packet that included information on the terrain, suggestions for modifications to equipment, a description of enemy weapons and tactics, and suggested countermeasures to the tactics. Such special South Vietnamese Army equipment as the M113 capstan recovery device was also described and detailed illustrations and explanations of South Vietnamese modifications to the M113, notably gunshields for the machine guns, were included.

Within three weeks after their arrival, the three armored units of the 25th Division participated in the first multibattalion armor operation to take place in Vietnam. Operation CIRCLE PINES was carried out in the jungle and rubber plantations twenty kilometers north of Saigon, where heavy growth favored the concealment of Viet Cong base camps. In places soft marshy soil impeded tank movement, but generally vegetation did not appreciably restrict either tanks or armored personnel carriers. By the close of this eight-day operation, more than fifty of the enemy had been killed and large amounts of arms and supplies had been captured. Of more importance was the fact that a large armored force had successfully invaded a Viet Cong jungle stronghold, forcing the enemy to move his base camps. The myth that armor could not be used in the jungle had been destroyed, and for that alone CIRCLE PINES will remain significant.

During April and May the 1st Battalion, 5th Infantry (Mechanized), continued operations around the Cu Chi region, returning to enemy bases in the Filhol Plantation and the Boi Loi and Ho Bo Woods. When conducting search and destroy operations in the base areas, mechanized infantry usually fought mounted, using the personnel carriers as assault vehicles. In the heavily wooded terrain the vehicles moved in column formation, breaking paths through the trees and thick brush and permitting the infantrymen to remain mounted and avoid booby traps. When the infantry located the Viet Cong the vehicles were moved rapidly toward enemy positions, with all guns firing in an attempt to overrun the enemy before the troops dismounted to make a thorough search. Using these techniques, in two months the battalion found and destroyed three base areas, killed 130 of the enemy, and captured thirty-six weapons.

In early June the 1st Battalion, 5th Infantry (Mechanized), began Operation MAKII, the first test of U.S. mechanized infantry's ability to operate in III Corps Tactical Zone during the wet season. This maneuver took the unit back into the Bao Trai–Duc Hoa area where it had fought in Operation HONOLULU in March, when the rice paddies were hard and the canals dry. The plan called for an immediate and rapid sweep of the respective company zones with the object of finding and destroying all enemy forces. To gain as much surprise as possible, Lieutenant Colonel Thomas U. Greer, the battalion commander, had the unit leave the base camp at Cu Chi just early enough to reach the line of departure by H-hour, which was 1100 on 3 June.

As the battalion reconnaissance platoon entered the zone of operation, it spotted and killed two Viet Cong soldiers. When the M113's churned over the dikes, more Viet Cong came out of the water where they were attempting to hide by submerging and breathing through hollow reeds. In a short time twelve of the enemy had been killed and nineteen captured. For the next three days the battalion searched the area, occasionally encountering a few Viet Cong soldiers. On 8 June, acting on information from prisoners, Company C discovered the second largest arms cache of the war to date—116 weapons and two tons of ammunition. The battalion learned from this operation that M113's could move through paddies covered with more than a foot of water but could not navigate damp, muddy paddies that had no standing water. The conclusion drawn from the maneuver was that most of the division area was suitable for mechanized operations, even during the rainy season.

Task Force Spur

Armored units arriving early in the Vietnam War literally had to invent tactics and techniques, and then convince the Army that they worked. While there was basic doctrine upon which to improvise, for Vietnam it needed expansion, modification, and, in some cases, combat testing. Not all innovations came from experienced armor leaders. Frequently, improvisation was necessitated by the tactical situation, but more often it came from the imagination of soldiers.

Several general officers advocated more armor. General Jonathan O. Seaman as commander of the 1st Division, and later as commander of II Field Force, constantly recommended the deployment of more armored units. Brigadier General Ellis W. Williamson asked for armor to support his 173d Airborne Brigade when the tanks of the 1st Squadron, 4th Cavalry, were being held at Phu Loi. General Fred C. Weyand, commander of the 25th Division, insisted on deploying armored units with his division. Brigadier General William E. DePuy, who as MACV J-3 was convinced that armored units could not be used, later changed his mind and as commander of the 1st Division constantly employed his armored units to seek out the enemy.

Not all early users of armor were generals. In early April 1966 Colonel Harold G. Moore, Jr., commander of the 3d Brigade, 1st

Cavalry Division (Airmobile), found a role for armored forces in the II Corps Tactical Zone near Chu Pong Mountain, twenty-seven kilometers west of Plei Me. The division was conducting Operation LINCOLN, and infantry units requested heavy artillery—175-mm. and 8-inch—to provide close support. The mission was given to Colonel Moore, who decided to move self-propelled artillery, escorted by Troop C, 3d Squadron, 4th Cavalry, through twenty kilometers of jungle where no roads existed. The armored column was to be guided and protected by elements of the 1st Squadron, 9th Cavalry (Air).

On 3 April Task Force SPUR, with nine M48A3 tanks, seventeen M113's, and the self-propelled artillery, struck out boldly from Plei Me into the jungle to the west. Tanks of the 3d Squadron, 4th Cavalry, guided by scouts of the 1st Squadron, 9th Cavalry, picked their way toward the planned artillery site with M113's, artillery, and cargo trucks following in their path. Task Force SPUR went through the twenty kilometers of jungle in seven hours, and for the next two days conducted search operations in the valley. When no Viet Cong were found, the armored unit returned to the artillery position to escort the guns back to Plei Me. The cavalrymen had covered over 108 kilometers of trackless jungle with the aid of air cavalry, demonstrating a far greater capacity for cross-country movement in the II Corps area than anyone had thought possible.

Battles on the Minh Thanh Road

In armored battles the mobility and heavy firepower of armored units often compensated for tactical mistakes. Some battles were extremely close, and caused changes to be made in operational procedures. One of these occurred in the III Corps Tactical Zone in 1966 as the 1st Infantry Division, probing into War Zone C, triggered a series of engagements on a dirt track called the Minh Thanh Road.

Operating with South Vietnamese forces, the U.S. 1st Infantry Division mounted a series of operations in eastern War Zone C during June and July 1966. The purpose was to open Route 13 from Saigon to Loc Ninh in Binh Long Province and to destroy elements of the 9th Viet Cong Division. The 9th was reported to be massing to seize the province capital of An Loc and several district capitals. By the end of Operation EL PASO II in early September 1966, five major engagements had been fought, and all three regiments of the 9th Viet Cong Division had withdrawn into sanctuaries deep in War Zone C along the Cambodian border, leaving behind some 850 dead. Highly effective counterambush tactics based on the firepower and mobility of armored forces were developed during three of the five engagements. These battles showed that armored cavalry with air and artillery support could more than hold its own against a numerically superior force, giving airmobile infantrymen time to join forces with the cavalry to defeat an enemy ambush.

The first of the three U.S. engagements took place when Troop A, 1st Squadron, 4th Cavalry, was ambushed by the 272d Viet Cong Regiment on 8 June 1966, north of the Ap Tau O bridge on National Highway 13. *(See Map 6, inset.)* The Viet Cong were deployed along a five-kilometer stretch of road in positions extending well beyond the length of the cavalry column. When the ambush was sprung most of the American troopers were able to reach a small clearing near the head of the column, where, with the help of artillery and air support, they desperately fought off the Viet Cong for four hours. When the battle ended, the enemy had lost over one hundred dead and four taken prisoner, as well as thirty individual and twelve crew-served weapons. Although successful the cavalry had made mistakes. Since original estimates of the enemy force were low, supporting fire was used primarily against the Viet Cong in the fighting positions near the cavalry force and other enemy forces were left free to maneuver. Although an infantry reaction force was committed toward the end, it did not arrive in time to be a decisive factor. After the commander and other principals had analyzed the battle, cavalry communications were changed and coordination of air and artillery was improved. Plans for reinforcement by airmobile infantry were developed to ensure quick arrival of reaction forces designed to fight off the main attack and to provide troops for blocking the enemy withdrawal routes.

Lessons learned on 8 June paid dividends on 30 June when the 271st Viet Cong Regiment attacked Troops B and C, 1st Squadron, 4th Cavalry, farther north along Highway 13 near Srok Dong. This time, when the ambush hit Troop B, Troop C rapidly maneuvered to reinforce. Coordination of fire support had vastly improved and tactical air and artillery were immediately and effectively employed. The relief force arrived in time to engage the Viet Cong before they could withdraw, while exploitation forces were inserted behind the enemy as far west as the Cambodian border, where another engagement took place. Enemy losses were heavy—270 dead, 7 taken prisoner; 23 crew-served weapons, and 40 small arms captured.

Encouraged by the two earlier successes of the 1st Squadron, 4th Cavalry, Major General William E. DePuy, 1st Infantry Division commander, directed development of a plan to lure the enemy into

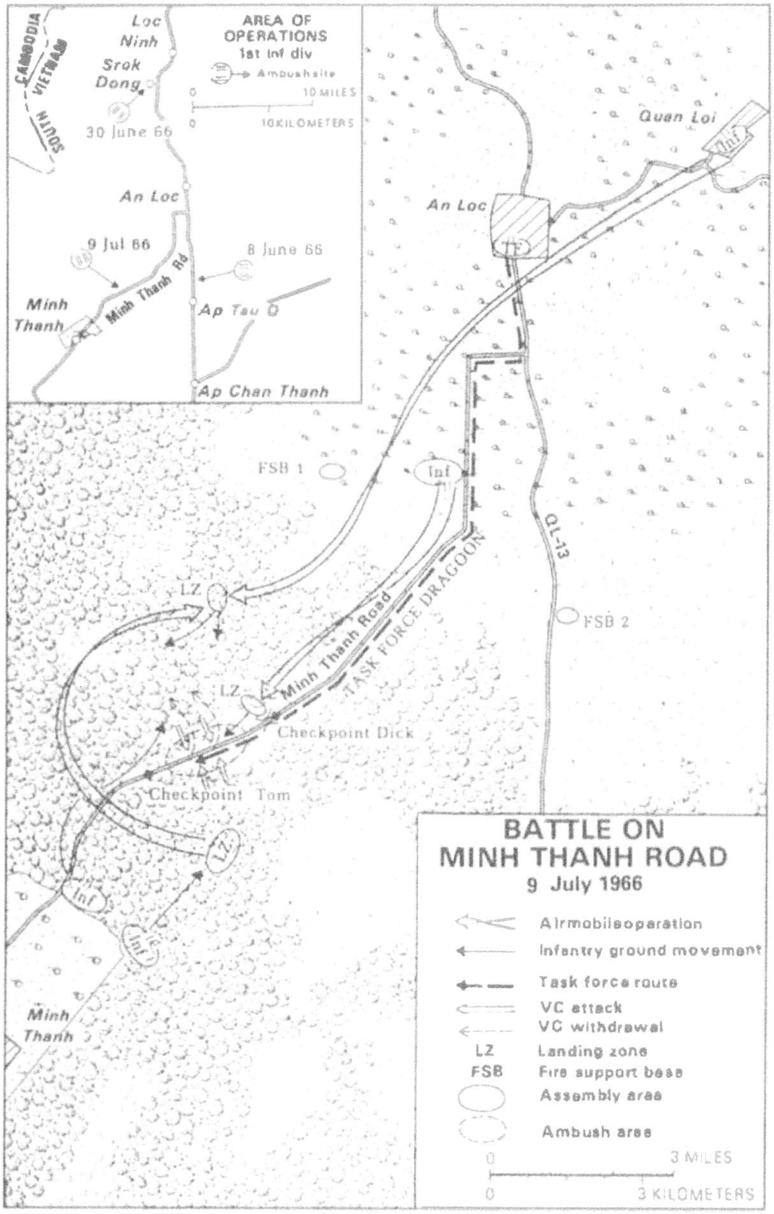
Loc Ninh
Srok Dong
30 June 66
CAMBODIA
SOUTH VIETNAM
AREA OF OPERATIONS
1st Inf div
Ambush site
0 10 MILES
0 10 KILOMETERS
An Loc
9 Jul 66
8 June 66
Minh Thanh Rd
Minh Thanh
Ap Tau O
Ap Chan Thanh
Quan Loi
Inf
An Loc
TF
FSB 1
Inf
QL-13
LZ
FSB 2
Minh Thanh Road
TASK FORCE DRAGOON
LZ
Checkpoint Dick
Checkpoint Tom
LZ
Inf
Inf
Minh Thanh
BATTLE ON
MINH THANH ROAD
9 July 1966
Airmobile operation
Infantry ground movement
Task force route
VC attack
VC withdrawal
LZ Landing zone
FSB Fire support base
Assembly area
Ambush area
0 3 MILES
0 3 KILOMETERS

MAP 6

attacking an armored cavalry column. Colonel Sidney B. Berry, Jr., 1st Brigade commander, prepared a two-phased flexible plan that could be easily modified for attacks on either Route 13 or the Minh Thanh Road. Five possible enemy ambush positions were selected during the planning, and, as it turned out, the site selected as the most likely was where the enemy struck. To increase the chances that the enemy would attack, rumors were circulated for the benefit of Viet Cong agents that a small armored column would escort an engineer bulldozer and several supply trucks from Minh Thanh to An Loc on 9 July. The true size of the force, called Task Force DRAGOON, was a well-kept secret; actually it was composed of Troops B and C, 1st Squadron, 4th Cavalry, and Company B, 1st Battalion, 2d Infantry.

Phase I, a deception plan involving an airmobile feint, covered the movement of artillery and supporting forces. The infantry forces that were to attack the flanks of the ambush and block withdrawal routes had to be in position to act quickly if the ambush occurred. The 1st Brigade began positioning these forces on 7 July, with infantry and artillery at Minh Thanh, more infantry and artillery just north of Minh Thanh Road, and an infantry battalion near An Loc. *(Map 6)* The trap was set.

At 0700 on 9 July 1966 Phase II started as the task force moved south from An Loc on National Highway 13, turned southwest on Minh Thanh Road, and arrived at Checkpoint Dick at 1025 without incident. There were artillery and air preparations on the western side of the bridge at Checkpoint Dick to soften up possible enemy concentrations. Following the air strikes, the 3d Platoon, Troop C, supported by covering fire, moved across the bridge with two engineer minesweeper demolition teams. A quick check was made for demolition charges and mines but no evidence of an enemy attempt to sabotage the bridge was found. Since the cavalrymen were now only 1,500 meters from the site selected earlier as the most likely ambush location, tension among the troops mounted. The Troop C commander directed the 1st Platoon to cross the bridge, pass through the 3d Platoon, and advance down the road toward Minh Thanh.

As the 1st Platoon moved past the 3d Platoon, a planned air strike was made near Checkpoint Tom while a CH-47, a helicopter with four .50-caliber machine guns, a 40-mm. grenade launcher, and two 7.62-mm. machine guns, struck the area southwest of Checkpoint Tom. There was no return fire. At 1100, midway between Checkpoints Dick and Tom, the crew of the lead vehicle of the 1st Platoon spotted ten Viet Cong running across the road. Minutes later when ten more crossed, the 1st Platoon's lead tank blasted them with canister. The tank fire brought an almost unbelievable volume of enemy fire on the entire Troop C column. The enemy had taken the bait.

At the beginning the 1st Platoon took the brunt of the enemy fire; the commander of the lead tank was killed. Within a few minutes the platoon leader reported his scout section out of action, and a little later he himself was wounded. As the platoon began to draw back under the heavy pressure, the platoon sergeant, who had taken command, moved to the front of the column to get the lead tank remanned and fighting. He directed the M132, a flame-thrower, to send liquid fire into the enemy positions on the north side of the road. Two of the 1st Platoon's M113's were hit and burst into flames. The 1st Platoon now had two tanks and four M113's firing at the enemy. The 2d Platoon, leading with an M48A3 tank, closed rapidly on the 1st Platoon and deployed in a herringbone formation, concentrating its fire to the north side of the road.[2] The 3d Platoon, heavily engaged as soon as the first rounds were fired, could not move forward to join the 1st Platoon and a 300-meter gap existed between the two platoons. The Viet Cong were unable to take advantage of the gap, however, because of the intense fire. Tracked vehicles along the entire column were firing as rapidly as possible, continuing to jockey for position and avoid the enemy antitank fire while artillery fire and air strikes hit the enemy positions.

The task force commander ordered Troop B forward to relieve the enemy pressure and called for more artillery and air support. At first the enemy's main attack had seemed to come from the south, but it was soon apparent that the enemy force was concentrated to the north side of the road. The plan for infantry reinforcement was put into action while the cavalrymen fought. When Troop B closed on the tail of Troop C, the fighting intensified. Within forty-five minutes the tanks had fired more than 50 percent of their canister and the M113's were nearly out of .50-caliber ammunition. Several Troop B tracked vehicles filled the 300-meter gap between the 2d and 3d Platoons of Troop C, and one platoon was assisting the lead element of Troop C.

With Troop B well disposed throughout the length of the Troop C column, the squadron commander ordered Troop C to

[2] The herringbone required all armored vehicles to stop immediately in an ambush, pivot to face outward at an angle, with alternate vehicles facing either side of the road, and fire all weapons. Softskin vehicles hid behind the armored vehicles. The maneuver was developed by the 1st Squadron, 4th Cavalry. It was a favorite of theirs and was used by many units.

pull back to Checkpoint Dick for resupply. Some supporting infantry were by then attacking the flanks of the ambush force while others were flown north in helicopters to take blocking positions. The battle raged for another half-hour, then the enemy began to leave protective cover and run away from the withering fire of the cavalrymen and supporting forces. As the Viet Cong fled, infantry, artillery, and tactical aircraft intercepted and destroyed them. An infantry sweep the following day discovered small groups of Viet Cong still trying to escape the trap. The searching forces found 240 of the enemy dead, took 8 prisoners, and captured 13 crew-served weapons and 41 small arms. By 1630 of 10 July the search was complete, and the 1st Squadron, 4th Cavalry, withdrew to Minh Thanh. The enemy plan to seize An Loc failed: the 1st Squadron, 4th Cavalry, had reopened Route 13, a vital line of communications, and had assisted in defeating the 9th Viet Cong Division.

Two significant facts emerge from these engagements. First, contrary to tradition, armored units were used as a fixing force, while airmobile infantry became the encircling maneuver element. Second, the armored force, led by tanks, had sufficient combat power to withstand the mass ambush until supporting artillery, air, and infantry could be brought in to destroy the enemy. Engagements with armored elements forcing or creating the fight and infantry reinforcing or encircling were typical of armor action in 1966 and 1967.

Armored forces, like other American units, generally avoided deliberate night actions in the early days of the Vietnam War. The scarcity of night fighting equipment, poor training of U.S. forces in night fighting, the difficulty of crashing through a dark jungle in armor at night, fear of ambush, and a general reluctance to fight at night, all militated against planned night actions. Armored operations at night were either reactions to enemy attacks or defenses of night positions. Such techniques as the use of helicopters and artillery flares for directing armored units and the employment of tank searchlights to illuminate likely ambush sites were eventually developed, but for most of the early years the night belonged to the enemy.

In an effort to change this situation armored leaders developed several techniques. One, nicknamed thunder run, involved the use of armored vehicles in all-night road marches using machine gun and main tank gun fire along the roadsides to trigger potential ambushes. While this procedure increased vehicle mileage and maintenance problems, it often succeeded in discouraging enemy road mining and ambushes. Highway 13 from Phu Cuong to Loc Ninh became known as Thunder Road because of the frequency of these runs and their similarity to those in the Robert Mitchum movie. Roadrunner operations, named after the cartoon character, although similar to the thunder runs were carried out by larger units on armed route reconnaissance that looked for trouble spots. These operations took place both day and night.

The Blackhorse Regiment

Although armored operations in Vietnam were catalysts for new concepts and innovations, there seemed to be, at MACV staff level, a lingering reluctance to deploy armored forces, especially those with M48A3 tanks. Nowhere is this better illustrated than in the events that preceded the arrival of the 11th Armored Cavalry Regiment, the Blackhorse Regiment.[3] Proposals were made to move the 11th Cavalry to Vietnam as early as December 1965, when General Westmoreland requested the regiment for the purpose of maintaining security along Route 1. His subsequent desire to use the unit for other missions precipitated a discussion of the regiment's table of organization and equipment. In late December 1965, the Military Assistance Command, Vietnam, requested equipment modifications to the 11th Cavalry tables, including substitution of light M41A3 tanks for medium tanks in the tank companies of the regiment's squadrons and M113's for both medium tanks and M114 reconnaissance vehicles in the cavalry platoons. After evaluating the proposed changes, the Department of the Army concluded that the regiment could not be sent as early as General Westmoreland had requested if all proposed changes were made.

The answer of the U.S. Military Assistance Command, Vietnam, that a mechanized brigade was required in lieu of the regiment, created considerable consternation among armor officers in the 11th Cavalry and in the Pentagon. It seemed that the largest armor unit yet requested for Vietnam would be eliminated before it had a chance to perform, and with it would go the hopes of many who believed that more armored forces were needed. The request for a brigade prompted a study by the Army staff, which considered as alternatives deploying a mechanized brigade, reshaping the 11th Cavalry, or sending the 11th Cavalry as it was then organized.[4]

3 Although General Westmoreland later stated that he welcomed with enthusiasm the arrival of the M48A3 medium tank in the 11th Cavalry, the facts bear out that not all the MACV staff shared his enthusiasm.

4 The Department of the Army action officer on this study was Lieutenant Colonel

Deployment of a modified 11th Armored Cavalry Regiment, with M113's substituted for the medium tanks and reconnaissance vehicles of the cavalry platoons, was considered the best alternative in view of the regiment's unusual capability for decentralized operations. The cavalry regiment had a higher density of automatic weapons, possessed long-range radios, and had more aircraft than a mechanized brigade. It had better means of gathering intelligence, was capable of rapid internal reinforcement, and possessed its own artillery in its squadron howitzer batteries.

When agreement on the unit's organization was reached, the 11th Cavalry began final preparations for Vietnam. Since M113's were to replace tanks in the cavalry platoons, they were modified for use as fighting vehicles by attaching a shield for the .50-caliber machine gun, and pedestals and shields for two side-mounted M60 machine guns. The concept and design were exactly that adopted by the South Vietnamese armor forces three years earlier, and subsequently recommended to American units by the advisers to the Vietnamese Armor Command. With the modifications the M113 was called the armored cavalry assault vehicle, or ACAV, a name coined by the 11th Cavalry troopers, probably in memory of the tanks the M113's replaced.

The 11th Cavalry arrived in Vietnam in early September 1966. Shortly after its arrival the Military Assistance Command welcomed its second U.S. Army tank battalion, the 2d Battalion, 34th Armor, commanded by Lieutenant Colonel Raymond L. Stailey. Part of the 4th Infantry Division before being sent to Vietnam, the battalion was attached to II Field Force in the III Corps Tactical Zone to replace the 1st Battalion, 69th Armor, which had moved to the II Corps area. On 19 September Company B was detached to the 1st Infantry Division at Phu Loi, and on 5 October Company A was detached to the 25th Infantry Division at Cu Chi. Finally, Company C was sent north to I Corps Tactical Zone until December. The practice of parceling out its tank companies was to haunt this battalion throughout its service in Vietnam; seldom did it have more than one tank company under battalion control. This unfortunate practice, so characteristic of the French in Indochina, was symptomatic of a command with few armored units. It reached a new high later in the war when, for a period of several months, the commander of the 2d Battalion, 34th Armor, again had no tank companies to command. The 11th Cavalry also suffered from the

George S. Patton, who would later command the 11th Cavalry in Vietnam. Feeling in the department was that if a compromise was not reached, neither the 11th Cavalry nor any other large armored unit would be deployed.

ACAV Prepares to Escort a Truck Convoy

detachment practice, and there were periods when the headquarters controlled only the regimental air cavalry troop.

After Operation Attleboro in September and October 1966, units of the 11th Cavalry returned to Bien Hoa to continue Operation Atlanta, whose mission was to clear and secure lines of communication in three provinces near Saigon and to secure the new Blackhorse Base Camp, 13 kilometers south of Xuan Loc. At first the operation was limited to clearing and securing Route 1 from Xuan Loc to Bien Hoa and Route 2 to the base camp. As Atlanta continued, however, the 11th Cavalry extended its operations away from the roads and throughout the area.

From the standpoint of the number of enemy killed, and, more important, from the number of roads opened to military and civilian traffic, Atlanta was a success. Regimental experience varied from roadrunner and convoy escort duties to cordon and search operations in which the squadrons sealed off an area and then worked, both mounted and dismounted, to drive out the enemy. Throughout, the regiment was able to move at will both on and off the roads, and experienced little difficulty with the terrain. Areas hitherto considered Viet Cong sanctuaries were entered by armored columns that destroyed base camps, fortifications, and supplies.

It was during Operation ATLANTA that the 11th Cavalry fought its first major battle. Twice the enemy tried to ambush and destroy resupply convoys escorted by units of the 1st Squadron, but in both attempts was defeated by the firepower and maneuverability of the cavalry. The second of these two ambushes took place on 2 December 1966 near Suoi Cat, fifty kilometers east of Saigon. The steps taken in this action illustrate a procedure for dealing with ambushes that became standard in the regiment.

When intelligence reports indicated that there was an enemy battalion in the vicinity of Suoi Cat, the 1st Squadron conducted a limited zone reconnaissance but found no signs of the enemy. Shortly thereafter, on 2 December 1966, Troop A was handling base camp security, Troop B was securing a rock quarry near Gia Ray, and the balance of the squadron was performing maintenance at Blackhorse Base Camp. *(Map 7)* Early that morning a resupply convoy from Troop B, consisting of two tanks, three ACAV's (modified M113's) and two 2½-ton trucks, had traveled the

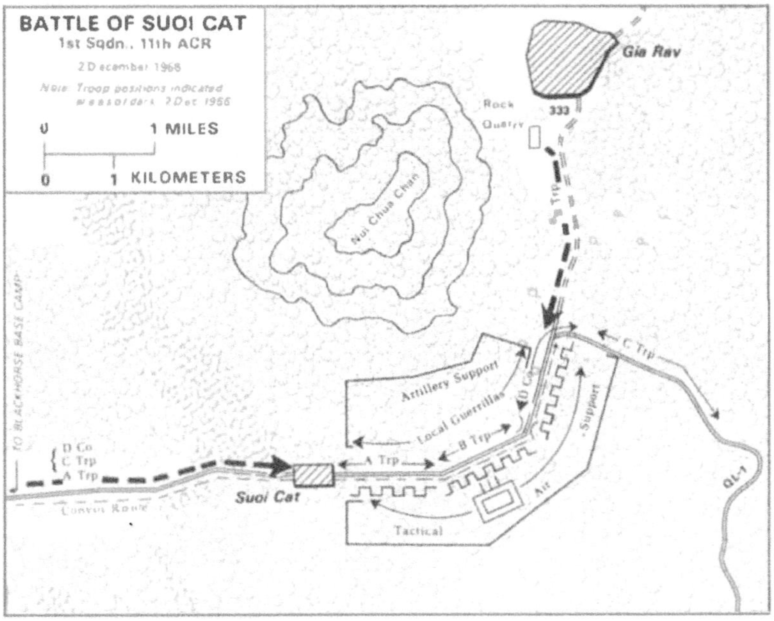

MAP 7

TANKS AND ACAV'S FORM DEFENSE PERIMETER AT BRIDGE SITE. *Distance between vehicles was much less than armor doctrine stated because of need for mutual support and to prevent infiltration.*

twenty-five kilometers from the rock quarry to Blackhorse without incident.

At 1600 the convoy commander, Lieutenant Wilbert Radosevich, readied his convoy for the return trip to Gia Ray. The column had a tank in the lead, followed by two ACAV's, two trucks, another ACAV, and, finally, the remaining tank. Lieutenant Radosevich was in the lead tank, and after making sure that he had contact with the forward air controller in an armed helicopter overhead, moved his convoy out toward Suoi Cat. As the convoy passed through Suoi Cat, the men in the column noticed an absence of children and an unusual stillness. Sensing danger, Lieutenant Radosevich was turning in the tank commander's hatch to observe closely both sides of the road when he accidently tripped the turret control handle. The turret moved suddenly to the right, evidently scaring the enemy into prematurely firing a command detonated mine approximately ten meters in front of the tank. Lieutenant Radosevich immediately shouted "Ambush! Ambush! Claymore Corner!" over the troop frequency [5] and led his convoy in a charge through what had become a hail of enemy fire while he blasted both sides of the road. Even as Lieutenant Radosevich charged, help was on the way. Troop B, nearest the scene, immediately headed toward the action. At squadron headquarters, Company D, a tank company, Troop C, and the howitzer battery hastened toward the ambush. Troop A, on perimeter security at the regimental base camp, followed as soon as it was released. The gunship on station immediately began delivering fire and called for additional assistance, while the forward air controller radioed for air support.

When the convoy reached the eastern edge of the ambush, one of the ACAV's, already hit three times, was struck again and caught fire. At this point Troop B arrived, moved into the ambush from the east, and immediately came under intense fire as the enemy maneuvered toward the burning ACAV. Troop B fought its way through the ambush, alternately employing the herringbone formation and moving west, and encountering the enemy in sizable groups.

Lieutenant Colonel Martin D. Howell, the squadron commander, arrived over the scene by helicopter ten minutes after the first fire. He immediately designated National Highway 1 a fire coordination line, and directed tactical aircraft to strike to the

[5] Claymore Corner was the unit nickname for this particular point on the road. Everyone in the squadron knew exactly where it was, even without a map.

east and south while artillery fired to the north and west. As Company D and Troop C reached Suoi Cat, he ordered them to begin firing as they left the east side of the village. The howitzer battery went into position in Suoi Cat. By this time Troop B had traversed the entire ambush area, turned around, and was making a second trip back toward the east. Company D and Troop C followed close behind, raking both sides of the road with fire as they moved. The tanks fired 90-mm. canister, mowing down the charging Viet Cong and destroying a 57-mm. recoilless rifle.[8] Midway through the ambush zone, Troop B halted in a herringbone formation, while Company D and Troop C continued to the east toward the junction of Route 333 and Route 1. Troop A, now to the west of the ambush, entered the area, surprised a scavenging party, and killed fifteen Viet Cong.

The squadron commander halted Troop A to the west of Troop B. Company D was turned around at the eastern side of the ambush and positioned to the east of Troop B. Troop C was sent southeast on Route 1 to trap enemy forces if they moved in that direction. As Troops A and B and Company D consolidated at the ambush site, enemy fire became intense around Troop B. The Viet Cong forces were soon caught in a deadly crossfire when the cavalry units converged. As darkness approached, the American troops prepared night defensive positions and artillery fire was shifted to the south to seal off enemy escape routes. A search of the battlefield the next morning revealed over 100 enemy dead. The toll, however, was heavier than that. Enemy documents captured in May 1967 recorded the loss of three Viet Cong battalion commanders and four company commanders in the Suoi Cat action.

The success of the tactics for countering ambushes developed during Atlanta resulted in their adoption as standard procedure for the future. The tactics called for the ambushed element to employ all its firepower to protect the escorted vehicles and fight clear of the enemy killing zone. Once clear, the cavalry would regroup and return to the killing zone. All available reinforcements would be rushed to the scene as rapidly as possible to attack the flanks of the ambush. Artillery and tactical air would be used to the maximum extent. This technique was used with success by the 11th Cavalry throughout its stay in Vietnam.

---

8 The tank commander who spotted the enemy crew setting up the recoilless rifle thought the men were from South Vietnamese forces who had come to help. When the muzzle was pointed right at him, he blew the weapon and crew apart at point-blank range with canister. Canister is a tank main gun round containing 1,280 shot or 5,600 to 10,000 darts. It is similar in action to a shotgun shell.

Mine and Countermine

Although tanks and ACAV's were effective against the enemy when he could be found, they were vulnerable to the explosive antivehicular mine. For example in June 1966, while moving back into the Boi Loi and Ho Bo woods in III Corps Tactical Zone, the 1st Battalion, 5th Infantry (Mechanized), lost fourteen M113's to mines in eight days of operation. Only eight M113's were eventually returned to service. In the period January–March 1967 on Highway 19 east of Pleiku in II Corps, the 1st Battalion, 69th Armor, found 115 mines: 27 were detected and disarmed, 88 exploded and damaged tanks. From June 1969 to June 1970, the 11th Cavalry encountered over 1,100 mines in the northern III Corps Tactical Zone. Only 60 percent were detected; the other 40 percent accounted for the loss of 352 combat vehicles.

Generally, tank hulls proved capable of absorbing the shock of a mine explosion, preventing serious injuries to the crews and damage to interior components. But when an APC hit a mine, particularly an APC that was gasoline-fueled, several crew members were usually wounded seriously or killed. Drivers were especially vulnerable, and crew members frequently rotated this dangerous job. For these reasons tanks normally led in clearing operations or reconnaissance in force. A study of the six-month period from November 1968 to May 1969 found that throughout Vietnam 73 percent of all tank losses and 77 percent of all armored personnel carrier losses were caused by mines. Another study conducted in December 1970 found that mines accounted for over 75 percent of all combat vehicles lost. This was not news to armor troopers.

In past wars countermine equipment had been chiefly designed to clear lanes through minefields where the mines were laid in patterns. In Vietnam, however, such minefields were never encountered; instead, the enemy planted mines at random on a massive scale. Antitank mines ranged from pressure-detonated to improvised mines, some as heavy as 250 pounds. The enemy also recovered unexploded artillery and mortar shells and aircraft bombs and rigged them with pressure-detonated or command-detonated fuzes. Mines were set on roads and off roads, in open field and dense jungle. There seemed to be no pattern that applied countrywide.

American units dealt with the mine problem by trying to prevent the enemy from laying mines, by trying to detect implanted mines, and by deliberately detonating mines—usually with a tank. Traditionally, countermine operations were efforts to detect mines

ENGINEER MINESWEEPING TEAM CLEARS HIGHWAY 13

after they were emplaced but in Vietnam, with no set battle lines, the enemy could be attacked as he attempted to lay mines. Ambush patrols were set up at likely enemy mining locations, and sensors were used to detect the people emplacing mines. The best way to defeat random mining was to kill the soldiers who were laying the mines or destroy the supply system that furnished the mines. Anything short of that was bound to be frustrating work with little promise of success.

Units like the 2d Battalion, 2d Infantry (Mechanized), used night roadrunner operations in an attempt to discourage or kill those placing mines along the road at night.[7] In addition to conducting runs at random intervals, the roadrunners called for planned artillery fire and reconnoitered by fire between friendly night defensive positions. The 11th Armored Cavalry employed thunder runs using tanks, and where possible fired harassing artillery fire on habitually mined sections of road. Other units made extensive studies of the tactical areas and developed mine incident charts. These studies pinpointed areas that were constant mine problems

[7] The 2d Battalion, 2d Infantry (M), was one of three infantry battalions that had deployed dismounted and was subsequently mechanized in Vietnam in early 1967. The other units were the 2d Battalion, 8th Infantry (M), and the 4th Battalion, 23d Infantry (M). Interestingly, the 2d Battalion, 2d Infantry, was a part of the 1st Infantry Division, which had been denied its mechanized battalions before deploying in 1965.

and invariably exposed three common factors: all the pinpointed spots were close to areas dominated by the local Viet Cong; all afforded the enemy good cover and routes in and out; and all had a high rate of mine incidents when armored units were present.

The information gathered from these studies indicated that the use of ambush patrols at night could be a valuable means of preventing mining operations, but it was limited, particularly in armored units, by the number of men that could be spared from other duties. Since armored units ranged over wide areas it was also impossible to study each area long enough to acquire sufficient information to act upon. Sensors, used in locations where there was repeated mining, were passive in nature but were responded to by artillery fire when activated. While their use seemed to reduce incidents, the precise effect was difficult to measure.

If the enemy could not be prevented from laying mines, the next step was to find the mines by some means other than running over them with vehicles. A mine sweeping team or troops familiar with an area could often visually locate mines. Informers who received on-the-spot cash payments and a degree of anonymity for themselves were a moderately reliable source of information. Metallic mine detectors and individual probing were useful but time consuming. On the whole, more road mines were spotted by alert armor crewmen than were found by mine detectors. Armored units were often the security element for clearance teams, and in most corps tactical zones had a daily mission of road clearance by probing and by using minesweepers and vehicles. Clearing units used one tank on the road and one on each shoulder; the tanks on the shoulders preceded the roadbound vehicle to destroy any wires to command-detonated mines in the road. The wheeled vehicles carefully followed in the tracks of the lead vehicle. Even fake mine-laying by the Viet Cong was successful since it also had to be checked. No system of mine detection was markedly effective, however, and losses occurred regularly in clearing operations.

Most armored vehicle crewmen took preventive measures to reduce mine injury to themselves and damage to their vehicles. The men always wore flak jackets and steel helmets. The floors of tracked vehicles were sometimes overlaid with sandbags, ration and ammunition boxes, or unusable flak jackets to prevent mine blast penetration.[8] In most units troops rode on top of the vehicles,

[8] Loading the APC with sandbags as a false floor helped, but often created other problems. When the 1st Squadron, 1st Cavalry, arrived, the cavalrymen put two layers of sandbags in all M113's, plus all the other equipment, troops, and ammunition. Within forty-five days, they broke fourteen transmissions from overloading. Most units used a layer or two of ammunition boxes.

feeling that it was better to get blown off the top than to be blown up inside. The Viet Cong countered by placing mines in trees. Some armor leaders even went so far as to have the crews of lead vehicles wear ear plugs to reduce ear damage when a mine was detonated. Tanks survived mine damage much better than M113's. To reduce mine damage to M113's, "belly armor" kits arrived in 1969. When this supplemental armor was applied to M113's and Sheridans, it protected them from mine blast rupture, saved many lives, and gave the crews added confidence, but it did not solve the mine problem.

As early as 1966 commanders in the field began to ask for better devices to deal with the mine danger. They were in particular need of a mine detector that could be mounted on a vehicle and that was capable of finding any type of mine, metallic or nonmetallic, no matter how fuzed. Finally, in 1969, the U.S. Army, Vietnam, asked the Mobility Equipment Research and Development Center to provide a device that could detect or destroy low-density mines on roads and that could move faster than a man carrying the portable mine detector then in use. The center's answer was the expendable mine roller, a mechanism to be mounted on and pushed in front of an M48 tank. The roller was tested at Fort Belvoir, Virginia, and delivered to Vietnam for combat evaluation in the fall of 1969.

Although the 11th Cavalry, which made the first test, felt strongly that the device would tie down a much needed vehicle, it fitted one tank with the roller and tested for over eighty kilometers, but no mines were found. Eventually the device was damaged when it was taken into the jungle, for which it was not intended. The regiment concluded that it was unsatisfactory, primarily because of its twenty-ton weight and maintenance requirements. Again in the fall of 1969 the 1st Brigade, 5th Infantry Division (Mechanized), tested the roller in Quang Tri near the Demilitarized Zone, where it proved unsuitable for the soft sandy soil of the region and was eventually ruined by a mine. The 4th Infantry Division made the third test of the mine roller, which was mounted on a combat engineer vehicle in lieu of a tank. In an experiment the roller detonated four mines and the 4th Division requested more rollers. Eventually twenty-seven were used in Vietnam. At the end of American participation in the war, the mine roller had not been fully accepted, and there was still need for a mine destroyer that would allow rapid movement.

Tank-Mounted Mine Roller Prepares to Clear Highway 19

CHAPTER IV

Combined Arms Operations

Armored units fighting in Vietnam by early 1967 included one armored cavalry regiment, six mechanized infantry battalions, four armored cavalry squadrons, two tank battalions, an air cavalry squadron, and five separate ground cavalry troops. By this time it was apparent that armored units of all types were proving far more useful in combat than had previously been thought possible. General Johnson, Army Chief of Staff, after discussions of the use of armor with General Westmoreland in the summer of 1966, directed the Army staff to determine whether a pattern of armored operations involving both tanks and armored personnel carriers had begun to emerge in Vietnam. The resulting staff study recommended an analysis by the U.S. Army Combat Developments Command for the purpose of suggesting modifications in unit organization, equipment, training, and deployment.

The MACOV Study

In August 1966 General Johnson approved plans for a study titled Mechanized and Armor Combat Operations in Vietnam (MACOV) under the direction of Major General Arthur L. West, Jr. Between January and March 1967 a group of over 100 American Army officers and civilian analysts examined the combat record of armored and mechanized forces in Vietnam, gathering and studying information gleaned from the battlefield. The group evaluated over 18,000 questionnaires, 2,000 reports, and 83 accounts of combat in which battalions and larger units had participated. Its report did not subscribe to the opinion that Vietnamese jungles and swamps would swallow up armored vehicles, but concluded that habitual use of armored vehicles against insurgents in jungles and swamps necessitated some changes in armor tactics. The group found that extensive use had been made of the armored personnel carrier M113, modified with weapons and gunshields to become a tanklike fighting vehicle that was known as the ACAV—armored cavalry assault vehicle. The study group also found that more often than not U.S. mechanized infantry fought mounted, employing armored personnel carriers as assault vehicles to close with and destroy the enemy, and that mounted troops generally suffered fewer and less serious casualties than foot soldiers. Contrary to established doctrine, armored units in Vietnam were being used to maintain pressure against the enemy in conjunction with envelopment by airmobile infantry. Moreover, tanks and APC's frequently preceded rather than followed dismounted infantry through the jungle, where they broke trail, destroyed antipersonnel mines, and disrupted enemy defenses. These findings revealed that some departures from armor doctrine had been taking place.

The study group noted several of the special advantages armor possessed in area warfare, described enemy tactics against armor, and listed types of armor missions. The group concluded that while tank and mechanized infantry units were playing a significant role in Vietnam, cavalry units, both ground and air, were essential elements to the important business of finding, pursuing, and destroying the enemy. Among its important recommended changes for armored and mechanized units was that organization be standardized for future armored forces being sent to Vietnam. This recommendation followed the discovery that, because of extensive and undisciplined modification of tables of organization and equipment, no two armored units in Vietnam were organized alike. Believing it impossible for the Department of the Army to support such a diverse force structure, the study group recommended that the Army strictly enforce conformity with modified standard tables of organization and equipment of units going to Vietnam.

Major findings of the study were described in a training manual, a training film, and an air cavalry text; all were given worldwide Army distribution. The air cavalry training text in particular was used for several years by air cavalry units and provided a much needed reference work to explain the air cavalry mission to ground commanders unfamiliar with the concept. It was also useful in training troops scheduled to deploy with air cavalry units.

The training manual's coverage was very broad, and when used correctly the manual was a "how to" book for armored units in Vietnam. Considering that most of the information had never been published in one book before, the manual was a landmark. General Westmoreland wrote the foreward and later commented that the study had prompted him to ask for more armored and mechanized units in troop requests. The text discussed impassable terrain and maps showed the areas that could be traversed in the wet and dry seasons. (*See Maps 2 and 3.*) In addition, it described the enemy and the frustrating nature of area warfare. Various battle formations and procedures such as herringbone, thunder run, and roadrunner were described in detail. The manual also discussed the cloverleaf, a maneuver particularly suited to armored units, mounted or dismounted, when they were making a rapid search of a large area.

The impact of the study was something less than many hoped for. The findings were not surprising to amored troops who had served in Vietnam but were regarded with a jaundiced eye by others who had not served there. Some data collectors believed that they were called upon to justify the existence of armor units already in Vietnam or scheduled to go there, but most members of the study group were able to put their task into perspective, and none expressed the justification for the study so well as one who said, "Although I did not doubt the value of armor in Vietnam, I was, myself, unable to recommend how much, of what type, and where it could be deployed. It would take a study like MACOV to provide a basis for these recommendations."

The bulk and security classification of the report prevented its widespread dissemination. In seven thick volumes, the official study was classified secret and was supported by six classified data supplements nearly as long as the report itself. Although its volume and classification were necessary, potential readers were overwhelmed. Only 300 copies were printed, and few remain in existence today. The publication of the unclassified training manual and film was an effort by the study group to gain wider circulation for the information.

At the U.S. Army Armor School and the Combat Developments Command Armor Agency at Fort Knox, changes in troop and equipment tables were enthusiastically endorsed, but doctrinal changes were rejected. While the report was clearly intended only to supplement worldwide armor doctrine, both the agency and the school argued that the new concepts were not applicable to armor combat in other parts of the world. Apparently those engaged in formulating doctrine were less concerned with the study group's conclusions, which were based on several years of combat experience in Vietnam, than they were with hypothetical situations in other parts of the world.

The training establishment under the Continental Army Command (CONARC) was unwilling to accept the study group's observations on the unprecedented role of M113's as assault vehicles in Vietnam. The command noted that the term "tanklike" was misleading and that adopting as doctrine the employment of mounted infantry in a cavalry role was neither feasible nor desirable. Justification for its position seemed to be couched in contradictory terms. While the command agreed that more Vietnam-

HERRINGBONE FORMATION ASSUMED BY 3D SQUADRON, 11TH ARMORED CAVALRY, DURING OPERATION CEDAR FALLS. *This formation gave vehicles best all-round firepower when they were ambushed in a restricted area.*

oriented training and doctrine were needed by deploying units, it refused to heed those findings of the study that were most attuned to the actual combat situation in Vietnam. The Continental Army Command decided to leave the matter to the interpretation of local commanders, although these were the same commanders who had told the study group that a revision of doctrine was needed to reflect actual combat experience in Vietnam.

The command also rejected the report's recommendation that the psychological effect of armored and mechanized units upon the enemy be exploited, stating that any further study of this matter would probably be superfluous. The implication was that the psychological advantage was not that great in the first place. "The Vietnamese people," stated the Continental Army Command, "know too well of the French Armored Mechanized Units' defeat at the hands of the Viet Minh and the destruction that can be inflicted on a tracked vehicle by one Viet Cong with a small amount

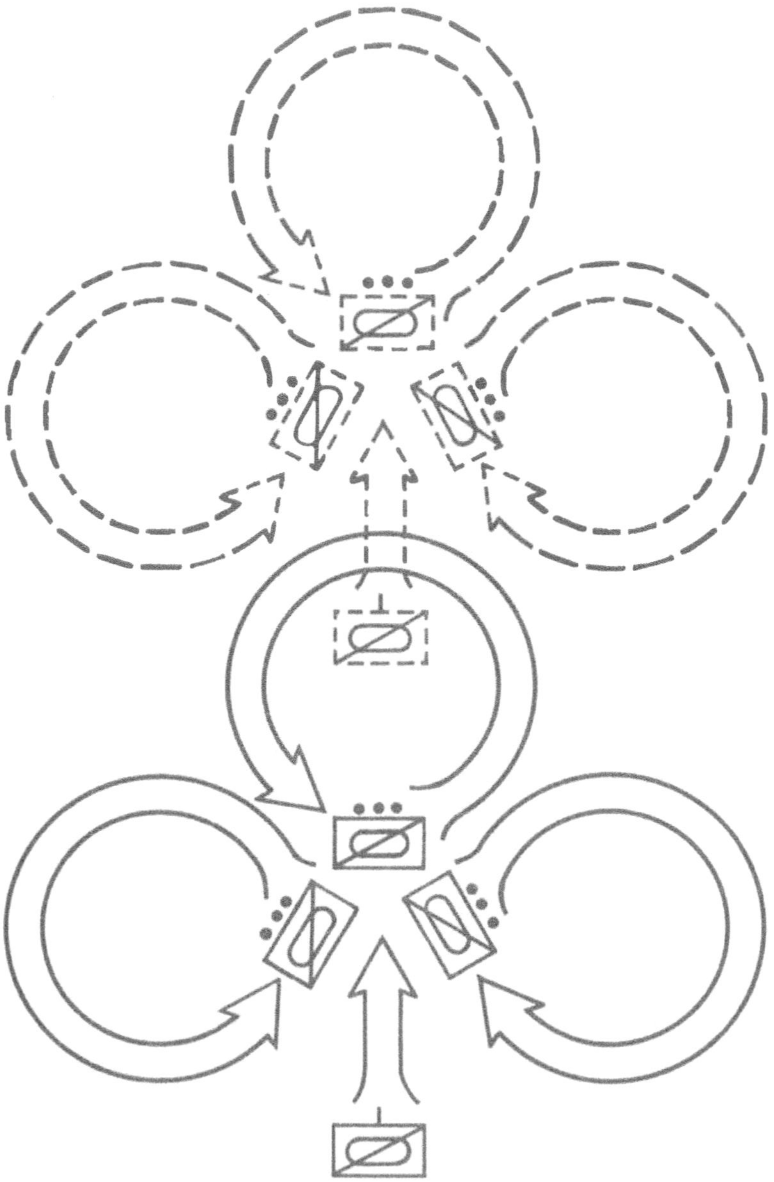

Diagram 6. Cloverleaf search technique used by armored cavalry troops.

of properly placed demolition material." To some elements within the command, the shock effect of armor, whether concrete or psychological, no longer existed, at least not in the case of the Viet Cong or the North Vietnamese Army.

The Army's equipment developer, the Army Material Command, and the doctrine agency, the Combat Developments Command, both commented on the report. The Army Material Command endorsed the majority of the study group's equipment recommendations but out of necessity qualified the approval with cost and time factors; it frequently noted feasibility but implied impracticality. The Combat Developments Command concluded that, with few exceptions, the recommendations should be carried out for Vietnam, and that certain of them were applicable to the Army worldwide.

The Combat Developments Command forwarded to the Department of the Army a strong endorsement of the study group's suggestion that increased emphasis be placed on the use of armored forces in warfare such as that in Vietnam. While the Army staff approved many specific recommendations, it did not agree that increased use of armored units in Vietnam was either necessary or desirable. In spite of the study group's observations on the usefulness of the M113 as an armored assault vehicle, the Army staff considered the results of such employment could only lead to a pyrrhic victory at best: "To modify and employ this means of transportation as an armored assault vehicle," it noted, "against an enemy who is daily improving the lethality and effectiveness of his armament not only decreases the capabilities for which the vehicle was originally designed, but can result in unnecessary friendly casualties." This position was totally inconsistent with the real world situation in 1967 in which U.S. and South Vietnamese armored forces were habitually and effectively employing their APC's and ACAV's as assault vehicles with great success.

One other circumstance worked against widespread acceptance of the recommendations: As the study group was preparing to leave for Vietnam in November 1966, Defense Secretary McNamara imposed an absolute troop ceiling on U.S. forces in Vietnam. This arbitrary ceiling was well below the total number already in the proposed troop program of the U.S. Military Assistance Command, Vietnam, and the United States Army, Vietnam. In other words, if more armored forces were wanted, other units had to be given up in order to get them. The situation was complicated by the fact that recommendations of an earlier study, Army Combat Operations in Vietnam, completed in 1966, had not yet been acted upon. The earlier recommendations dealt with infantry problems in Vietnam in the same detail as the armor study dealt with armor problems; they also required trade-offs, most of which had not yet been decided upon.

The armor study group applied itself to this problem in a straightforward way by incorporating the infantry study recommendations and summing up the cumulative effect of both studies. The group then selected some 4,000 troop spaces in the proposed force for the U.S. Military Assistance Command, Vietnam, that could be traded to carry out the combined recommendations of both studies. Since the spaces came largely from combat support troops, the logistics and administrative community vehemently denounced the trade-off, thereby heightening the opposition to the study group's recommendations.

The armor study raised again all the historic arguments for and against the use of armored forces in Vietnam. It provided a documented basis for discussion and, in fact, influenced the training and employment of many armored units. The Department of the Army subsequently approved organizational and equipment changes incorporating some of the recommendations. The armored cavalry assault vehicle, for example, is in the Army today, and fighting mounted from armored vehicles is an accepted practice. Vietnam related training at the Armor School was increased from two to twenty hours in mid-1967, although academic department heads expressed concern that the Army would overemphasize Vietnam at the expense of conventional armor employment. This attitude was in striking contrast to that of junior officers and students who knew they were destined for duty in Vietnam.

Armored units scheduled for Vietnam used the armor study group's training manual as a guide, but copies were difficult to obtain; many armor officers never saw it. Only a few years later, units and service schools were hard put to find the copies they had received.[1] Perhaps the most effective dissemination of the study findings came through the efforts of the group members, some of whom wrote service school lesson plans, contributed articles to periodicals, and made changes in units in which they served. All in all, the armor study accelerated changes in the theory of using armored forces that would be tested and validated by the battles of the *Tet* offensive of 1968.

---

1 In 1973 neither the Armor Agency nor the Armor School at Fort Knox had a copy of the study report and only a few copies of the manual could be found.

Cedar Falls—Junction City

Early combat operations in 1967 that were observed, recorded, and analyzed by General West's study group reflected a definite change in strategy for American and other free world forces in Vietnam. Until late 1966 General Westmoreland had employed "fire brigade tactics," reacting to enemy initiatives with his limited troop resources. By 1967 the buildup of U.S. forces permitted him some flexibility, and increases in tactical mobility improved the effectiveness of the reaction forces. Thus, in 1967 the mission of American and other free world units changed to one of offensive action against the main force enemy units. South Vietnamese forces were to be employed primarily in pacification. The initiative was passing to the free world forces.

In the III Corps Tactical Zone the first deliberately planned multidivision operation, CEDAR FALLS, was begun by II Field Force, Vietnam. The target was an extensive enemy base and logistical center that because of its geographical shape and strong defense was known as the Iron Triangle. (*Map 8*) This heavily jungled area, twenty-five kilometers north of Saigon, was an important center for the launching of enemy guerrilla and terrorist operations; it was frequently referred to as "a dagger pointed at Saigon." The plan was to seal the area, split it in half, thoroughly search it, and destroy all base camps and enemy forces.

The Iron Triangle was sealed by U.S. armored and airmobile units. The 2d Battalion, 22d Infantry (Mechanized); 1st Battalion, 5th Infantry (Mechanized); 2d Battalion, 34th Armor, and Troop B, 3d Squadron, 4th Cavalry, established blocking positions west of the Saigon River; the 1st Squadron, 4th Cavalry, E Troop, 17th Cavalry, and Company D, 16th Armor, were employed east of the Thi Tinh River. Having moved to an assembly area the day before, the 11th Armored Cavalry, less its 1st Squadron, attacked west from Ben Cat on 9 January to divide the area in two. Throughout the operation, units combed the Iron Triangle, uncovering base camps, food, equipment, and ammunition. Fighting was light and generally limited to scattered encounters with platoon-size or smaller groups. The value of CEDAR FALLS does not lie in the number of enemy casualties it produced but in the 500,000 pages of enemy documents it captured. These exposed the command structure and battle plans of the entire Viet Cong and North Vietnamese Army hierarchy. General Seaman described the operation as the largest intelligence breakthrough in the war.

Tracked vehicles, which had little difficulty in traversing the terrain, were assisted in the search by bulldozers. A task force of

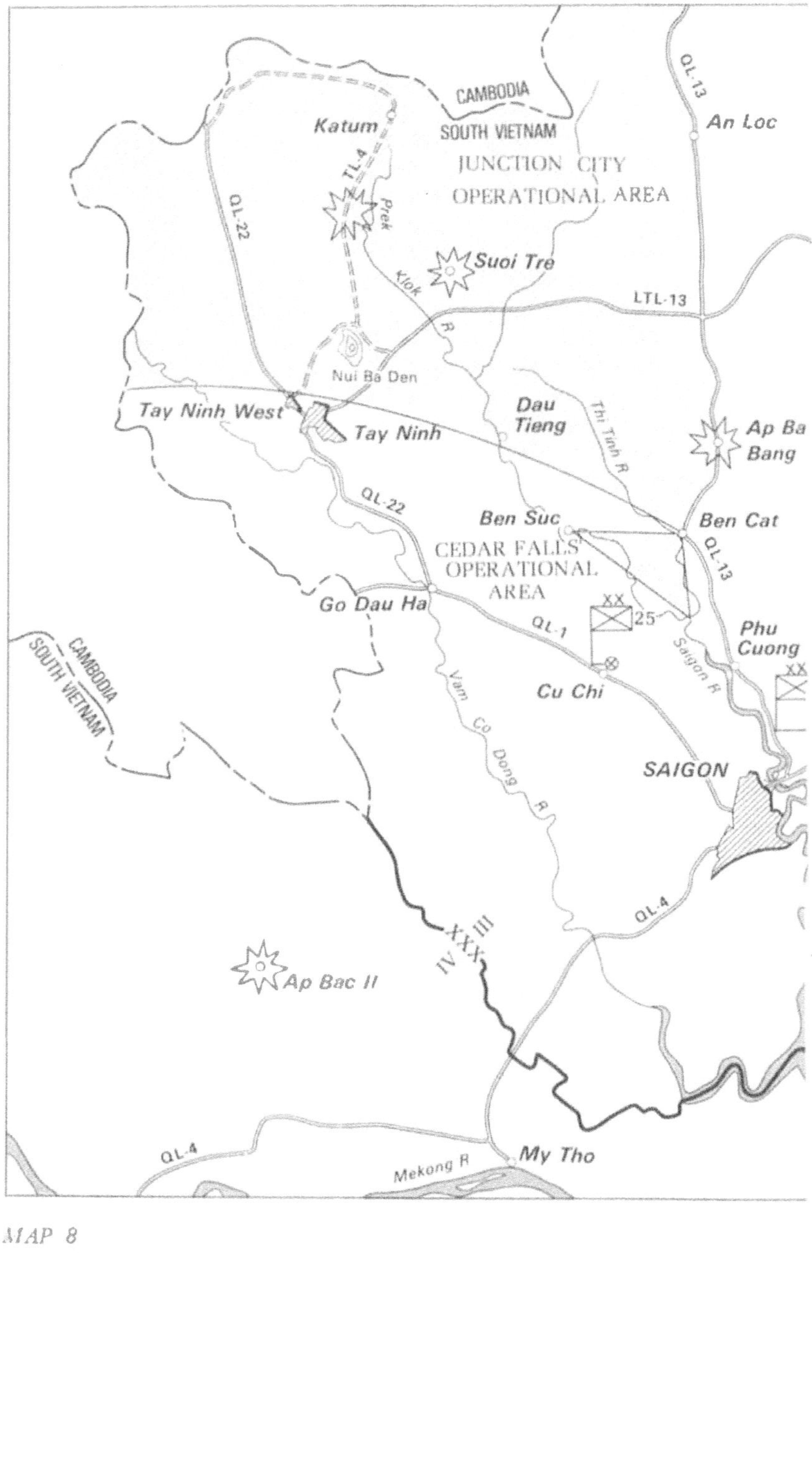
CAMBODIA
SOUTH VIETNAM
Katum
TL-4
JUNCTION CITY
OPERATIONAL AREA
QL-13
An Loc
QL-22
Prek
Suoi Tre
Klok
R
LTL-13
Nui Ba Den
Tay Ninh West
Tay Ninh
Dau
Tieng
Thi Tinh R
Ap Ba
Bang
QL-22
Ben Suc
Ben Cat
QL-13
CEDAR FALLS
OPERATIONAL
AREA
XX
25
Go Dau Ha
QL-1
Phu
Cuong
CAMBODIA
SOUTH VIETNAM
Vam
Co
Dong
R
Saigon R
Cu Chi
XX
SAIGON
QL-4
III
XXX
IV
Ap Bac II
QL-4
Mekong R
My Tho

MAP 8

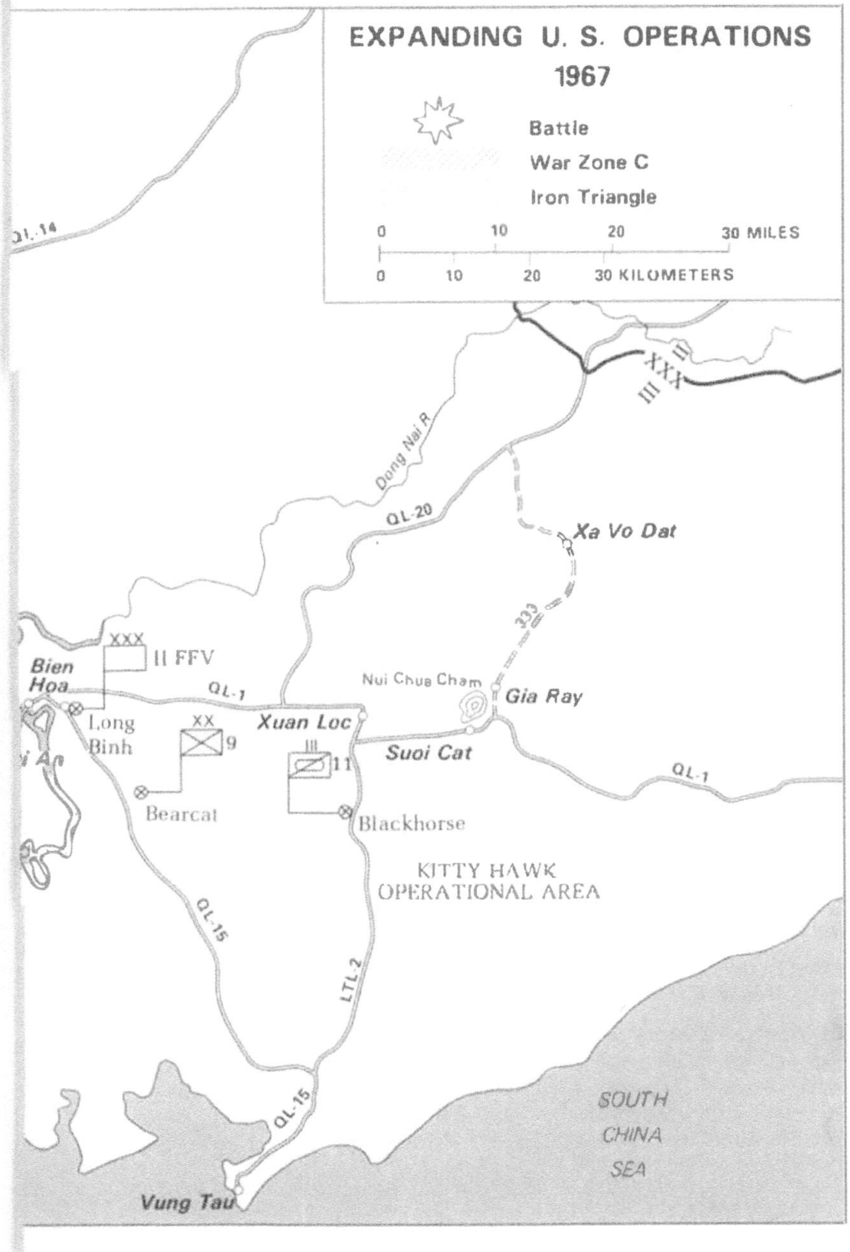
EXPANDING U. S. OPERATIONS
1967
Battle
War Zone C
Iron Triangle
0 10 20 30 MILES
0 10 20 30 KILOMETERS
QL-14
Dong Nai R
QL-20
Xa Vo Dat
333
XXX
II FFV
Bien Hoa
QL-1
Long Binh
XX
9
Xuan Loc
Nui Chua Cham
Gia Ray
Suoi Cat
III
11
QL-1
Bearcat
Blackhorse
KITTY HAWK
OPERATIONAL AREA
QL-15
LTL-2
QL-15
SOUTH
CHINA
SEA
Vung Tau

M113's AND M48A3 TANKS DEPLOY BETWEEN JUNGLE AND RUBBER PLANTATION IN OPERATION CEDAR FALLS

fifty-four bulldozers with four Rome plows and some tanks with dozer blades cleared more than nine square miles of jungle, and frequently led armored columns. In an interesting innovation, the 2d Battalion, 34th Armor, used tank-mounted searchlights to detect Viet Cong night movements along the Saigon River. Several successful night ambushes were conducted by directing tank fire against the enemy river traffic.

The wisdom of the 1966 decision to increase the number of mechanized infantry battalions from two to six was attested to by Brigadier General Richard T. Knowles, Commanding General, 196th Light Infantry Brigade, in a statement concerning the role of mechanized infantry in CEDAR FALLS.

Mechanized infantry has proven to be highly successful in search and destroy operations. With their capability for rapid reaction and firepower, a mechanical battalion can effectively control twice as much terrain as an infantry battalion. Rapid penetrations into VC areas to secure LZs for airmobile units provide an added security measure for aircraft as well as personnel when introducing units into the combat zone. The constant movement of mechanized units back and forth through an area keeps the VC moving and creates targets for friendly ambushes, artillery and air.

The 11th Armored Cavalry Regiment's success in CEDAR FALLS clearly confirmed the soundness of the unit's organization. The regimental commander, Colonel William W. Cobb, reported that the first rapid maneuver into the area and its accomplishment of search and destroy, screening, blocking, and security missions demonstrated the flexibility of his unit. He further stated:

> The search and destroy portion of Operation CEDAR FALLS was the final combat test of the modified TOE designed to tailor the regiment's organization to the requirements of the counterinsurgency operations in Vietnam. The search and destroy operations, plus the allied saturation and sniper patrols, and tunnel search operations proved the validity of the MTOE. There proved to be sufficient personnel in the basic maneuver element—the Armored Cavalry Platoon—to allow for required dismounted tunnel and patrolling operations while maintaining sufficient crew members on the ACAV's to maintain the platoon's mounted combat capabilities.

When Operation CEDAR FALLS ended on 25 January 1967, armored forces of II Field Force, Vietnam, were committed to Operation JUNCTION CITY, the largest operation of the war to that date. JUNCTION CITY was designed to disrupt the Viet Cong Central Office for South Vietnam (COSVN), destroy Viet Cong and North Vietnamese forces, and clear the War Zone C base areas in northern Tay Ninh Province, eighty-five kilometers northwest of Saigon. During the three-phase operation, armored units served in road security and search and clear operations and acted as convoy escorts and reaction forces.

Phase I, 22 February–17 March, consisted of establishing a horseshoe blocking position in northwestern War Zone C, then attacking into the open end of the horseshoe toward the *U* end of the position. From Fire Support Base I at 0600 on 22 February, a 1st Squadron, 4th Cavalry, task force began a twenty-kilometer move north on Provincial Route 4; its mission was to reinforce quickly the airborne and airmobile assault elements at the north end of the horseshoe. To send the cavalry along an uncleared route was a calculated risk, prompted by the hope that the enemy would not employ mines on one of his few partially paved supply routes. The gamble worked. The 1st Squadron raced unimpeded to reinforcing positions south of Katum, and the landings went without incident. As the cavalry moved north, the 2d Battalion, 2d Infantry (Mechanized), followed with artillery and engineering units to establish Fire Support Bases II and III.

At dawn on 23 February, the 2d Brigade, 25th Infantry Division, and the 11th Armored Cavalry Regiment began sweeping north between the sides of the horseshoe. There was scattered fighting as the armored units found base camps, hospitals, bunker systems, and small groups of Viet Cong. Mines and booby traps slowed the attack, and in the center of the horseshoe dense jungle made movement difficult. After reaching the northern limit of advance, the 11th Armored Cavalry Regiment pivoted and swept west with the other forces. Sporadic fighting continued.

For armored units, JUNCTION CITY was a task force operation. Combined arms operations at battalion and squadron level were normal, and mobility was stressed. Armored task forces with attached elements of infantry, artillery, tank, and cavalry roamed through the operational area. Infantry rode on the tracked vehicles and went into action as tank-infantry teams.

Although enemy resistance gradually stiffened throughout the area, the armored task forces finally drew out the elusive Viet Cong on the periphery of the operation. The mobile blocking forces were interfering with Viet Cong supply operations, and the enemy fought back. The resistance was particularly evident along Routes 4 and 13 as the enemy shifted eastward to avoid the JUNCTION CITY attacks. In this area three armored battles took place, each illustrating a different type of combined arms action. The first, at Prek Klok II, stressed firepower; the second, at Ban Bang, demonstrated mobility and staying power; and the third, at Suoi Tre, emphasized mobility and shock action.

While mines were not encountered in the first thrust north on Provincial Route 4, as operations progressed the enemy began to mine the road, hoping to cut the American force's primary resupply route. From random sniping and mining the enemy went to mortar attacks and night probes of fire bases. On the evening of 10 March, Fire Support Base II at Prek Klok II was defended by Lieutenant Colonel Edward J. Collins's 2d Battalion, 2d Infantry (Mechanized), which was minus its Company B, some engineer troops, and two batteries of 105-mm. artillery. The base straddled an airfield that the engineers were constructing. Tracked vehicles were placed around the perimeter at 50-meter intervals, and artillerymen, engineers, and infantrymen manned foxholes between the APC's. About 2030 a listening post sighted three Viet Cong and immediately pulled back into the perimeter; the base went to 75 percent alert status. An unearthly silence fell. Some thirty minutes later it was broken by the dull thump of enemy mortars firing. In a matter of seconds the entire area erupted with explosions as the enemy poured over 200 rounds of mortar and recoilless rifle fire into the base. When the barrage ended, Colonel Collins ordered the defenders to conduct a reconnaissance by fire of the area 200 to 600 meters beyond the perimeter.[2]

When the U.S. machine guns fell silent at 2220, the enemy launched a two-battalion ground attack from the east. The first wave of the assault came within hand grenade range, and the perimeter was enveloped in fire as the defenders answered with vehicle-mounted and ground machine guns, small arms, and artillery fire. Intense Viet Cong antitank fire, rocket propelled grenades and recoilless rifles, was directed against the APC's. Although the vehicles were positioned behind a low berm, three were struck by rocket grenades and one received a direct hit from a mortar round.

To support the main attack, smaller enemy forays were launched from the northeast and southeast. Trip flares and listening posts had been set out about fifty meters in those areas. According to Specialist 4 Thomas Lark, when the listening posts were brought in after the mortar attack, "We opened up on the VC when they hit our trip flares and after that we never had any trouble with the VC getting close to our perimeter." A secondary attack was also launched from the southwest, but here the enemy had to cross 500 meters of open ground. Amid explosions of recoilless rifle rounds, the defenders held their positions, pouring machine gun and small arms fire into the attackers. This secondary attack never gained momentum, although heavy enemy fire continued from the wooded area beyond the clearing. Supported by air strikes, artillery, and machine gun fire from "Spooky" (a C-47 aircraft with multi-barrel machine guns), the defenders repelled the brunt of the attack by 2300.

The battle of Prek Klok II was one-sided, for the enemy lost almost 200 men while the defenders lost three. The enemy had hoped to achieve a quick victory to bolster his sagging fortunes. Instead, a combined arms team of artillerymen, mechanized infantrymen and aircraft, using selective firepower and properly prepared defensive positions, had dealt him a severe defeat.

As JUNCTION CITY continued into Phase II, the enemy lost enormous amounts of supplies and was denied use of his vital communications centers. In an attempt to ease the pressure, the Viet Cong launched a desperate attack on 19 and 20 March against a fire base protected by the cavalry. The base, sixty kilometers north of Saigon near Ap Bau Bang on QL-13, was in flat country

[2] A reconnaissance by fire, or "mad minute" was used to disrupt enemy action. All weapons fired at maximum rate directly out of a perimeter on all sides until given a cease-fire order. The volume of fire and sound from an armored unit was awesome, and usually discouraged further enemy activity.

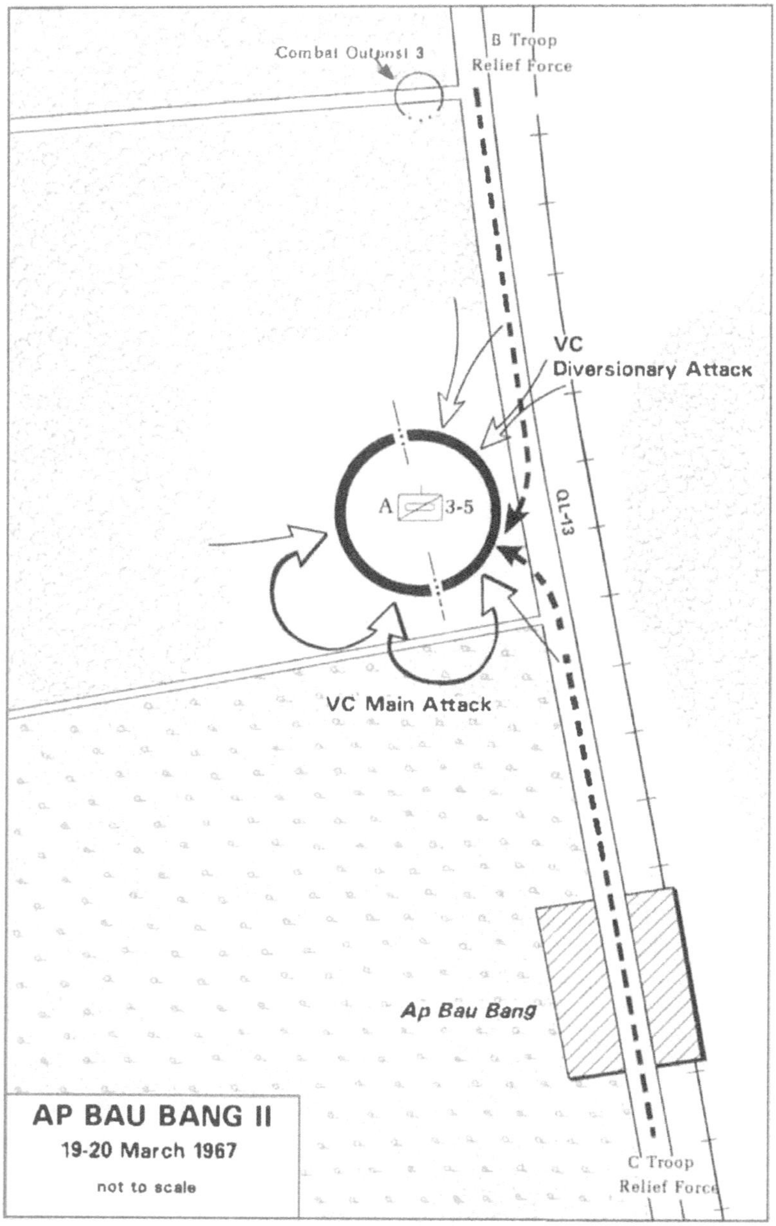
Combat Outpost 3
B Troop
Relief Force
VC
Diversionary Attack
A 3-5
QL-13
VC Main Attack
Ap Bau Bang
C Troop
Relief Force
AP BAU BANG II
19-20 March 1967
not to scale

MAP 9

with wooded areas to the north and west and a rubber plantation to the south. (*Map 9*) It was protected by Troop A, 3d Squadron, 5th Cavalry Regiment, and contained Battery B, 7th Battalion, 9th Artillery (105-mm.). One platoon from the troop occupied Combat Outpost 3, approximately 2,800 meters north. Troop B of the 3d Squadron was located to the north, and Troop C and the headquarters troop were protecting the squadron command post to the south.

At 2300 the northeast section of the fire support base was raked by fire from an enemy machine gun, but the gun was quickly silenced by return fire from tanks and armored cavalry assault vehicles. Shortly after midnight the base came under heavy fire from rocket grenades, mortars, and recoilless rifles, followed by a massed ground assault. Main attacks from the southwest and southeast were supported by a diversionary attack from the northeast. Troop A defenders at first held their own but requested that a ready reserve force be designated for use if needed. The 1st Platoon of Troop B, to the north, and the 3rd Platoon of Troop C, in the south, were alerted to assist.

The battle intensified as enemy troops reached the vehicles on the southwest portion of the perimeter, but with the help of more than 2,500 rounds of sustained artillery fire from all calibers of weapons the cavalry held. At times enemy soldiers were blasted off ACAV's by 90-mm. canister fire from nearby tanks. When the tanks ran out of canister, they fired high explosive rounds set on delayed fuses into the ground in front of the enemy. The result was a ricochet round that exploded overhead and showered fragments over the enemy units — a very effective weapon. Several defending vehicles were hit and destroyed by rocket grenade fire, and the gaps created in the line finally forced the troop to fall back to tighten its perimeter.

At 0115 the squadron commander, Lieutenant Colonel Sidney S. Haszard, gave Troop A permission to move its 2d Platoon from Combat Outpost 3 to the fire base, and ordered Troops B and C into action. The 2d Platoon had to attack the enemy in order to get through to the defenders. When the 3d Platoon of Troop C arrived, the Troop A commander directed that it sweep south of the perimeter along the tree line. Continually firing its weapons, the platoon swept the southwest side of the fire base, then doubled back and entered the defensive line from the southeast. At the same time, while en route to the base at thirty miles an hour, the 1st Platoon of Troop B literally ran over a hastily set ambush. Just as the platoon arrived, the enemy launched another attack. The Troop A commander directed it to sweep the entire perimeter. Circling around the outside of the base with headlights and searchlights flashing and weapons firing, the platoon crushed the attack.

The enemy's next attack, at 0300 from the south, was easily repelled by the five cavalry platoons in the base and air support that eventually totaled eighty-seven sorties through the night. Troop A then conducted a series of counterattacks, clearing an area 800 meters deep around the perimeter and reducing the enemy fire. At 0500 under illumination from flares and searchlights, the enemy could be seen massing for another attack from the south and southeast. Tactical aircraft and artillery were quickly employed and the attack never gained momentum. Although sporadic enemy fire continued, the six-hour battle ended, leaving over 200 enemy dead on the battfield and three American soldiers killed.

The success of the defense hinged on the mobility of the armored units, the heavy firepower—artillery and air support—and the tactics used. The armored vehicles had not been dug in and were not fenced in with wire. Throughout the attacks, ACAV's and tanks continuously moved backward and forward, often for more than twenty meters, to confuse enemy gunners and meet attacks head on. The movement added to the shock effect of the vehicles, for none of the enemy wanted to be run over. In addition, reinforcing platoons all carried extra ammunition on their vehicles and provided resuppy during the battle.

The last major armored action in JUNCTION CITY occurred only a day after the Ap Bau Bang fight, when the enemy launched an unprecedented daylight attack against Fire Support Base Gold near Suoi Tre. The fire base was occupied by the 3d Battalion, 22d Infantry (–) and the 2d Battalion, 77th Artillery (–). The 2d Battalion, 12th Infantry, the 2d Battalion, 34th Armor, and the 2d Battalion, 22d Infantry (Mechanized), were conducting search and destroy operations nearby.

The 2d Battalion, 34th Armor, commanded by Lieutenant Colonel Raymond Stailey, moved north on 20 March, led by Company A, 2d Battalion, 22d Infantry, which had been sent to link up with the tanks. By nightfall the two battalions had joined and set up camp within two kilometers of each other. Earlier that afternoon the Scout platoon of the mechanized battalion had cleared a trail about 1,500 meters to the north but had been unable to locate a ford across the Suoi Samat. Lieutenant Colonel Ralph W. Julian, commander of the 2d Battalion, 22d Infantry, decided that the next day his units would move north on the trail, then swing east to search for a ford across the upper reaches of the stream.

On the opposite side of the Suoi Samat, about two kilometers northeast of the tank battalion's position, infantrymen and artillerymen were improving perimeter defenses at Fire Support Base Gold. The next morning at 0630, an ambush patrol from the 3d Battalion, 22d Infantry, engaged a large force of Viet Cong moving toward the base and at the same time the base came under heavy mortar attack. Over 600 rounds pounded the camp as waves of Viet Cong emerged from the jungle, firing recoilless rifles, rocket grenades, automatic weapons, and small arms. The ambush patrol was quickly overrun and was unable to return to the base. As the fighting grew more intense, the armored units to the south were ordered across the Suoi Samat to reinforce the embattled fire base. Colonel Julian immediately moved part of Company C and an attached tank platoon north on the trail cleared earlier by the Scout platoon.

While the remainder of the column closed, conditions worsened at the fire base. Colonel Marshall Garth, the brigade commander, said "If a vehicle throws a track, leave it. Let's get in there and relieve the force." Personnel carriers in the lead straddled each other's paths in order to clear a trail wide enough for tanks, while lead elements using compasses continued their search to the east in an attempt to find the Suoi Samat and a ford.

At Fire Base Gold, counter fire was seeking out the enemy mortars that were pounding the defenders. The enemy concentrated against the east side of the perimeter until, at 0711, Company B reported that its 1st Platoon had been overrun. A reserve force of artillerymen helped to reestablish the perimeter, but forty-five minutes later the enemy had again broken through the 1st Platoon. Within a few minutes, positions on the northeastern portion of the Company B perimeter were completely overrun by a human wave attack. Company A sent a force with desperately needed ammunition to assist Company B. Then, on the northern perimeter, the Viet Cong swarmed over a quadruple .50-caliber and attempted to turn it on the defenders, but the weapon was blown apart by the artillery. To make matters worse, Company A reported penetrations in portions of its northern perimeter.

The urgency of the situation was again conveyed to Colonel Julian by Colonel Garth's order that the stream was to be crossed "even if you have to fill it up with your own vehicles and drive across them." Following instructions from a helicopter overhead, the armored column finally crossed the stream and moved toward the fire base. To the northwest the 2d Battalion, 12th Infantry, advancing on foot, had reached the defenders. From the air, Colonel Julian directed Lieutenant Colonel Joe Elliot, Commander of the

2d Battalion, to secure the western sector of the fire base. The mechanized forces were ordered to enter just south of the 2d Battalion, 12th Infantry, and swing around the perimeter, consolidating the remainder.

On the smoke-covered battlefield the reinforced defenders were still in desperate straits. Artillerymen were firing beehive rounds, steel flechettes released at the muzzle of the weapon. When the supply of beehive was exhausted, they switched to high-explosive direct fire at point-blank ranges. The eastern sector of the perimeter had fallen back under heavy pressure to positions around the artillery pieces. The Viet Cong were within five meters of the battalion aid station and within hand grenade range of the command post.

Into this chaos came the tanks and APC's, crashing through the last few trees into the clearing. The noise was overwhelming as the new arrivals opened up with more than 200 machine guns and 90-mm. tank guns. The ground shook as tracked vehicles moved around the perimeter throwing up a wall of fire to their outside flank. They cut through the advancing Viet Cong, crushing many of them under the tracks. The Viet Cong, realizing that they could not outrun the encircling vehicles, charged them and attempted to climb aboard but were quickly cut down. Even the tank recovery vehicle of Company A, 2d Battalion, 34th Armor, smashed through the trees with its machine gun chattering. Most of the crew, who were all mechanics, were throwing grenades, but one calm mechanic sat serenely atop the vehicle, his movie camera grinding away.

Relief was evident in the faces of the defenders as tracked vehicles quickly tied in with the 2d Battalion, 12th Infantry. "It was," exulted Lieutenant Colonel John A. Bender, the fire base commander, "just like the late show on TV, the U.S. Cavalry came riding to the rescue." Master Sergeant Andrew Hunter recalled, "They haven't made the word to describe what we thought when we saw those tanks and armored personnel carriers. It was de-vine." With victory almost within grasp of the enemy, the tanks and APC's had turned the tide. When the smoke cleared, it was apparent that the enemy had not only been defeated but had lost more than 600 men.

JUNCTION CITY II ended on 15 April as the enemy faded away. Armored units played a major role in JUNCTION CITY and proved that in most areas of War Zone C, a cavalry squadron or mechanized infantry battalion could more effectively control a large area than any other type of unit. Although routes over the difficult terrain had to be carefully selected, tracked units moved through most of the dense jungled area. Tanks were invaluable in breaking trails through seemingly impenetrable vegetation. The ability of armored forces to move rapidly and to arrive at the critical place with great firepower gave them a significant advantage.

Mechanized Operations in the Mekong Delta

The extensive rice paddies and mangrove swamps of the canal-laced delta were very different from the jungled areas of Operation CEDAR FALLS-JUNCTION CITY in III Corps Tactical Zone. But in the delta, with few high elevations, M113's could move as freely as rivers and major canals permitted. The 2d Battalion, 47th Infantry (Mechanized), and the 5th Battalion, 60th Infantry (Mechanized), of the U.S. 9th Infantry Division—two armored units employed in this region—conducted successful combined American-Vietnamese operations throughout 1967. Typical missions included reconnaissance in force, route and convoy security, night roadrunner operations, cordon and search of villages, and rapid reinforcement.

Flooded rice paddies slowed, but did not prevent cross-country movement. Small canals up to three meters in width were crossed with balk bridging. In the case of larger canals and rivers, which were major obstacles because their banks were usually steep or composed of loose soil, bulldozers or explosives were used to construct entry and exit routes. Mechanized units quickly discovered that when track shrouds were removed to prevent the buildup of mud between track and hull the M113's swimming ability was impaired. Navy landing craft were therefore required for transportation across major rivers and canals. Route reconnaissance by air was always important but was essential during the monsoon season.

The delta's open, level terrain permitted ground troops to engage with organic weapons at much greater range than that of the point-blank fighting normal in the jungle. One of the hardest battles fought by mechanized infantry in the delta occurred at the village of Ap Bac II on 2 May 1967. Ap Bac II was a base area for the 514th Viet Cong Battalion, and the enemy pattern of movement between base areas had suggested the probability of the battalion's presence near Ap Bac II on 2 May.

The original plan of the 2d Brigade, 9th Infantry Division, was to conduct an airmobile search and destroy operation with two battalions of infantry. On 2 May, however, when no helicopters were available the insertion of a blocking force was deleted from the plan. Movement of two battalions abreast without a blocking force in the rear was regarded by many as "forcing toothpaste

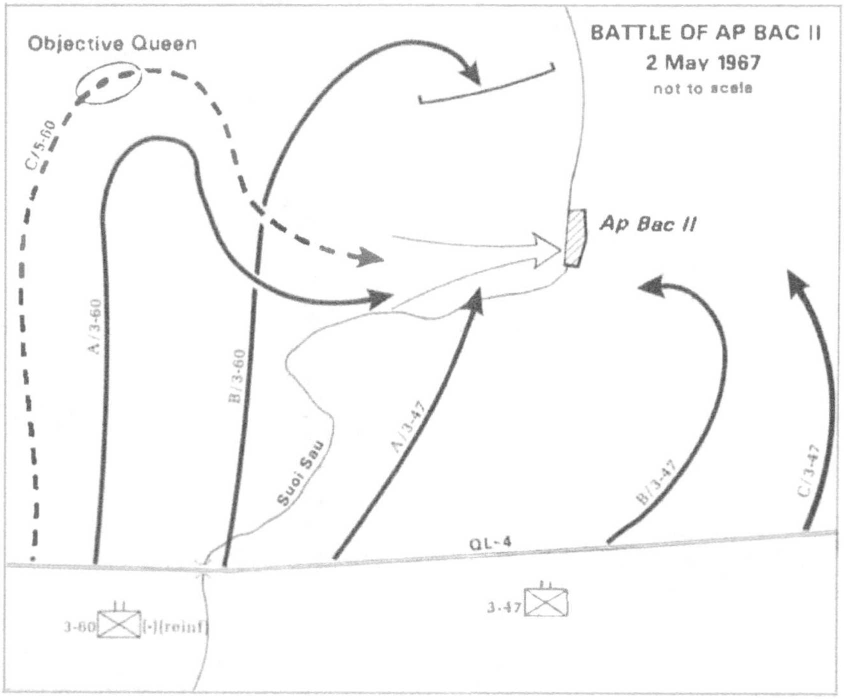

MAP 10

from a tube," and there appeared little likelihood of a significant encounter. Company C, 5th Battalion, 60th Infantry (Mechanized), manned the left flank under the control of the 3d Battalion, 60th Infantry. (*Map 10*) First Lieutenant Larry Garner's mechanized company was given the deeper objective because its mobility would permit a quick search of the area. It was hoped that tracked vehicles could make up for lack of a blocking force.

By 0830 the M113's of Company C were advancing north, crossing paddies surrounded by narrow, earthen dikes. Mostly dry, the paddies easily supported tracks, but crossing the many canals and streams proved more difficult. Company C found none of the enemy during its northward sweep; however, to the east, Company A, 3d Battalion, 47th Infantry, encountered stiff resistance as it approached the Suoi Sau. The steep banks of the stream were dotted with thatched huts and lined with dense vegetation. A squad, maneuvering across the stream, was quickly pinned down by heavy automatic weapons fire. Within minutes all who had crossed the stream had been hit. Two companies of the 3d Battalion, 47th Infantry, moved in on the right of Company A, while Company B, 3d Battalion, 60th Infantry, moved to block the northern escape route. At 1300, with blocking forces in a reversed "C," Company C of the 5th Battalion and Company A of the 3d Battalion of the 60th Infantry were ordered east to fill the open end of the blocking positions. The eleven M113's of Company C had to maneuver through inundated areas that appeared impassable. Crossing two fairly deep streams, the company chose routes that brought it abreast of Company A, 3d Battalion, 60th Infantry, on a 1,000-meter assault line by approximately 1530. Under cover of artillery and air bombardment, the companies crossed more irrigation ditches and by 1700 were poised for the attack.

On order, artillery fire stopped and the tracked vehicles surged forward, while blocking units supported by fire. The mechanized company moved rapidly across the open rice paddy, its machine guns searching out the enemy bunkers along the wood line. At the woods infantrymen dismounted and attacked the enemy soldiers who had been pinned down by heavy fire. Although stunned by the shock of the assault, the Viet Cong continued to resist, and the infantry was forced to move among the bunkers destroying the enemy with grenades.

Company A, moving on foot to the right of Company C, met heavy resistance and finally stalled about 100 meters from the bunker line. The company commander requested help from Company C, which responded by moving four M113's to aid the dismounted attack. Since darkness had set in, further reinforcement was considered impractical and the units on hand had to finish the job. Additional fire support by the M113's, a charge by the attacking companies, and heavy fire superiority finally broke the enemy's defense. The companies pressed the attack, forcing the Viet Cong from their bunkers and annihilating those who tried to escape. A sweep of the battle area early the next morning indicated that the enemy had lost the equivalent of a reinforced company. Two U.S. soldiers had died.

Colonel William B. Fulton, the brigade commander, noted that the speed, shock effect, and heavy firepower provided by the personnel carriers, along with supporting artillery, had kept the enemy soldiers in their bunkers until the infantry was literally on top of them. Lieutenant Colonel Edwin W. Chamberlain, Jr., commander of the 3d Battalion, 60th Infantry, stated that since the tracked vehicles proved capable of negotiating more terrain than had been thought possible, there should always be an initial attempt at

TANKS AND ACAV'S SECURE SUPPLY ROUTES IN 25TH INFANTRY DIVISION AREA. *Sandbags modify vehicles for security role as movable pillboxes.*

mounted movement in order to capitalize on the additional firepower of the vehicular-mounted machine guns.

Mechanized infantry units in the delta were extremely flexible and were used alternately in mechanized, airmobile, and dismounted infantry operations. First and foremost, however, they were mechanized infantry, capitalizing on their vehicular mobility to close with the enemy, then dismounting and assaulting, supported by a base of fire from the vehicles. This is exactly what Company C had done.

Route Security and Convoy Escort

The missions universally shared by armored units throughout Vietnam were furnishing route security and convoy escort. Few tasks were more important than keeping the roads safe and protecting the vehicles, men, and supplies that used them. At the same time, no task was more disliked by armored soldiers. When it was done correctly it could be boring, tedious, and in the minds of many, a waste of time and armored vehicles. When it was done poorly, or when the enemy was determined to oppose it, it was dangerous, disorganized, and, again in the minds of many, a one-way ride to disaster.

General Westmoreland's directive had called for opening the roads, making them safe, and using them. Carrying out the order was a different problem in each area. In one instance in mid-1966 the task became an intricate, large-scale operation that led to battles along Highway 13 and the Minh Thanh Road involving the 1st Squadron, 4th Cavalry. In another situation in the highlands, a significant part of the 4th Infantry Division's armored forces—at first the 1st Squadron, 10th Cavalry, and the 1st Battalion, 69th Armor, and after September 1967 the 2d Squadron, 1st Cavalry—continuously secured roads throughout the division's area of operation in the II Corps Tactical Zone.[3] In the first three months of 1967, almost 8,000 vehicles a month under armored protection traversed Highway 19 to Pleiku without incident.

The primary route security technique used in the highlands was to establish strongpoints along the road at critical locations, and each morning have a mounted unit sweep a designated portion of the route. The unit then returned to the strongpoint where it remained on alert, ready to deal with any enemy action in its sector. When forces were insufficient to man strongpoints twenty-four hours a day, each convoy using the road was provided with an escort force, a measure that caused heavy wear on the armored vehicles. Securing roads by using static positions had the disadvantages that the Viet Cong quickly noted them and mined all logical vehicle positions with the result that the protective force soon lost vehicles in the strongpoints. When the 2d Squadron, 1st Cavalry, was attached to the 4th Infantry Division, the division abandoned the strongpoint system in favor of offensive patrolling missions several thousand meters from main routes, a tactic that made a much more effective use of armor.[4]

Sometimes the security and escort missions were given operational names and continued for six months or more. One such operation, KITTY HAWK in the III Corps Tactical Zone, required a cavalry squadron to secure the Blackhorse Base Camp and the Gia Ray rock quarry, to escort convoys, and to conduct local reconnaissance in force. (*See Map 8.*) During 1967 the 3d Squadron, 5th

---

3 The 1st Squadron, 10th Cavalry, which had arrived in August 1966, would spend most of its five-year effort in Vietnam repetitiously securing the road net in the Pleiku area. The 2d Battalion, 8th Infantry, which was mechanized in 1967, occasionally shared this mission.

4 The 1st and 2d Squadrons, 1st Cavalry, both arrived in August 1967 with the diesel powered M113A1, modified with ACAV kits. Since these squadrons were never relieved of their respective assignments to the 1st and 2d Armored Divisions, personnel from these units became the first and only armored crewmen since World War II authorized to wear an armored division combat patch on the right shoulder of their uniforms.

Cavalry, and a squadron from the 11th Armored Cavalry Regiment alternated in this job. Missions such as this were required throughout Vietnam because of constant enemy threats.

Occasionally an escort or security mission was not successful, and usually intensive after action investigation revealed that the unit had been careless. Such was the case with a platoon of Troop K, 11th Armored Cavalry Regiment, in May 1967. When the smoke cleared from a well-planned Viet Cong ambush, the platoon had paid a heavy price: seven ACAV's had each been hit 10 times by antitank weapons and the lone tank had 14 hits. Of forty-four men in the convoy, nearly half were killed and the remainder wounded. Investigation revealed that the road had been cleared that morning by a responsible unit, but the fact that an ambush was set up later proved that it was dangerous to assume that one pass along a road cleared it of enemy forces. In this case there were further errors of omission. No planned platoon action was put into effect when the enemy attacked; no command and control alternatives were provided in the event of a loss of radio communication; no signals or checks were in effect to alert troop headquarters to the platoon's plight; no artillery or air support was planned for the route of march. The lesson from this disaster was that no mission should be considered routine.

Disasters were uncommon to road security missions, but much could be learned from them. On one occasion the law of averages, troop turnover, and the boredom of a routine task caught up with the 3d Squadron, 5th Cavalry, while it was on road security. This incident in late December 1967 illustrates how overconfidence, poor planning, and lack of fire support could combine to strip the cavalry of its inherent advantages. On 22 December the squadron was to assume responsibility for Operation KITTY HAWK. The squadron staff prepared its plans for convoy escort, with convoys scheduled to move on 27 and 31 December. At the last moment, the 3d Squadron's assumption of the KITTY HAWK mission was delayed until 28 December and the 11th Armored Cavalry Regiment performed the escort duty on 27 December. The two-day delay caused the staff of the 3d Squadron, 5th Cavalry, to be less attentive to the second convoy escort mission on the 31st.

On 28 December the 3d Squadron moved into Blackhorse Base Camp, and the next day the squadron operations officer was reminded of the responsibility for the escort on 31 December. Mission requirements were discussed over insecure telephone lines by the staffs of the squadron and the 9th Infantry Division, and were then passed to Troop C, which had the mission. The squadron daily staff briefing on 30 December did not include a discussion of the escort mission and the squadron commander remained unaware of it. The Troop C commander, familiar with the area, believed the sector to be relatively quiet, a fatal assumption because combat operations had not been conducted in the area for over thirty days. He planned a routine tactical road march to Vung Tau, sixty kilometers to the south, to rendezvous with the convoy at 0900 on 31 December. Two platoon-size elements were to make the march while the troop commander remained at Blackhorse with the third platoon, ready to assist if needed. The platoons were to leave Blackhorse at 0330 on 31 December, moving south on Route 2. One platoon was to stop along Route 2, about a third of the way to Vung Tau, and spend the night running the road back to Blackhorse to prevent enemy interference on the route. The other platoon was to continue to Vung Tau, pick up the convoy, and escort it to Blackhorse. The convoy would be rejoined en route by the platoon conducting roadrunner operations.

The column moved out on time to meet the convoy. The lead platoon, commanded by the 2d Platoon leader, consisted of one tank from the 3d Platoon, two ACAV's from the 2d, and the troop command and maintenance vehicles employed as ACAV's. The next platoon, commanded by the 3d Platoon leader, consisted of one tank from the 2d Platoon, two ACAV's from the 3d Platoon, two from the 1st Platoon, and the 1st Platoon's mortar carrier minus its mortar. The tanks, each leading a platoon, intermittently used driving lights and searchlights to illuminate and observe along the sides of the road.

About nine kilometers south of Blackhorse, Route 2 crested a slight rise, ran straight south for two kilometers, and then crested another rise. The sides of the road had been cleared out to about 100 meters. As the lead tank started up the southernmost rise at 0410, the last vehicle in the convoy, the mortar carrier, was leveling off on the straight stretch two kilometers behind. Suddenly a rocket propelled grenade round hit the lead tank, killing the driver and stopping the tank in the middle of the road. An ambush then erupted along the entire two-kilometer stretch of road. A hail of grenades quickly set the remaining vehicles of the lead platoon afire; intense small arms fire killed most of the men riding atop the vehicles. As the trailing platoon leader directed his platoon into a herringbone formation, the mortar carrier was hit by a command detonated mine, exploding mortar ammunition and destroying the carrier. The tank with the last platoon was hit by a rocket grenade round, ran off the road, blew up, and burned. The surprise was so

OH–6A Observation Helicopter and Two AH–1G Cobras En Route to Conduct Visual Reconnaissance Near Phuoc Vinh

complete that no organized fire was returned. When individual vehicles attempted to return fire, the enemy, from positions in a deadfall some fifteen meters off the road, concentrated on that one vehicle until it stopped firing. Within ten minutes the fight was over.

At daybreak on the last day of 1967, the devastating results of the ambush were apparent in the battered and burned hulks that lay scattered along the road. Of eleven vehicles, four ACAV's and one tank were destroyed, three ACAV's and one tank severely damaged. The two platoons suffered 42 casualties; apparently none of the enemy was killed or wounded. This costly action showed what could happen on a routine mission in South Vietnam. Indifference to unit integrity, breaches of communication security beforehand, lack of planned fire support, and wide gaps between the vehicles stacked the deck in the enemy's favor. Charged with guarding a convoy, the unit leader failed to appreciate his own unit's vulnerability.

Elsewhere in the III Corps Tactical Zone other U.S. units were performing similiar route and convoy security missions. The 3d Squadron, 4th Cavalry, 25th Infantry Division, operated almost continuously along Route 1 from Saigon to Tay Ninh. The squadron's air cavalry troop worked with it, providing first and last light reconnaissance along main routes. By mid-1967 the squadron was escorting an average of 8,000 vehicles per month. In late summer it began so-called night thrust missions sending out mock convoy escorts at night to test the enemy reaction. After a month-long test without significant enemy action, the 3d Squadron, 4th Cavalry, began escorting night logistical convoys from Saigon to Tay Ninh, a mission that continued through 1967.

Air Cavalry Operations

As a necessary complement to ground armored forces, air cavalry units brought a new dimension to the Vietnam conflict. The first air cavalry unit, the 1st Squadron, 9th Cavalry, 1st Cavalry Division, exploited the concept and literally wrote the book on air cavalry operations. Few other air cavalry units, particularly those with divisional cavalry squadrons, were assigned air cavalry roles at first; instead, they were used to escort airmobile operations—like armed helicopter companies. After the armor study and the assignment of more experienced and innovative commanders, air cavalry troops finally began to operate in air cavalry missions.[5] Quite often, however, rotation of commanders, particularly senior commanders, required that lessons be relearned time and again. There was thus a continuing discussion on the proper role and the command of air cavalry units.

In units that properly used air cavalry, operations followed a daily pattern. Upon receipt of information indicating enemy activity, an air reconnaissance was conducted by the troop to determine whether or not further exploitation by ground forces was required. If ground reconnaissance was desirable, the troop commander usually committed his aerorifle platoon. A standby reserve force could be called by the troop commander if the situation required. The air cavalry troop commander controlled all reaction forces until more than one company from a supporting unit was committed. At that time control passed to the commander responsible for the area, and the operation was conducted like a typical ground or airmobile engagement, often with the air cavalry remaining in support. Major General John J. Tolson, Commander, 1st Cavalry

[5] The air cavalry troop of the 3d Squadron, 4th Cavalry, experienced difficulty on arrival in 1966, when the 25th Division Aviation Battalion tried to have it converted to an aviation gunship unit in the Aviation Battalion. The conversion was finally rejected, but the air cavalry used aviation company tactics until September 1967, when it assumed normal scout operations and employed low-level flying techniques.

Division, clearly stated his feelings about air cavalry: "I cannot emphasize how valuable this unit [1st Squadron, 9th Cavalry] has been to me as division commander. Over 50 percent of our major contacts have been initiated by actions of this squadron."

To be successful, air cavalry operations had to be swiftly followed by ground or airmobile elements of the division or regiment. Unfortunately in many units with air cavalry, reported leads were frequently not followed up. Consequently, many air cavalry units adopted the unofficial practice of developing leads that could be handled by the air cavalry itself.

In October 1967 two air cavalry squadrons, the 3d and 7th Squadrons, 17th Cavalry, arrived. Because they were the first units of their type to be assigned to U.S. infantry divisions in Vietnam, their integration into the force was accomplished with some difficulty. Most problems reflected a lack of knowledge on the part of the division commander and staff concerning capabilities, limitations, and basic support needs of air cavalry squadrons. There was an unfortunate tendency to use the aircraft for command and control and for transportation in airmobile operations rather than for reconnaissance. At the outset, therefore, air cavalry was not used to best advantage, and there was some misuse. Only after commanders became more aware of the capabilities of their air cavalry squadrons was proper employment achieved, and in some cases the process was slow and painful.[6]

Other Free World Armor

As early as April 1965, discussions had been held on the deployment to Vietnam of armored units from other nations. Surprisingly, the impetus for such discussion came from Ambassador Maxwell Taylor, and centered on the proposed use of a tank company from New Zealand. At the Honolulu Conference in April 1965 this proposal was disapproved, but by September when the Australian task force arrived an armored personnel carrier troop was included. Equivalent to a reinforced American platoon, the troop was quickly put to work with the U.S. 173d Airborne Brigade. In October free world armor strength increased when the Republic of the Philippines sent a security force of seventeen APC's and two M41 tanks.

The Royal Thai Army forces that arrived in Vietnam in 1967 brought with them an M113 platoon and a cavalry reconnaissance

[6] Air cavalry development and employment is fully treated in a companion monograph of the Vietnam Studies, *Airmobility, 1961–1971*, by Lieutenant General John J. Tolson. Although air cavalry is a component of mounted warfare, space limitations prevent the detailed treatment of it here.

TROOPS OF 1ST AUSTRALIAN ARMOR REGIMENT IN FRONT OF AUSTRALIAN CENTURION TANK *receive briefing at Vung Tau.*

troop. By 1969 this force had been increased to three cavalry troops and a total of over 660 armor soldiers. The Koreans asked permission to deploy a tank battalion, but the request was disapproved in midsummer 1965 on the grounds that the area was inappropriate for tanks. Later, Korean and American tank-infantry operations in the area enabled the Koreans to acquire APC's on permanent loan from the United States, and these were employed as ACAV's. Finally, in 1968, the Australians sent twenty-six Centurion tanks and an additional cavalry platoon. The Centurions, the only tanks other than U.S. tanks used in Vietnam by the free world forces, had 84-mm. guns and successfully operated east of Saigon near Vung Tau.

Although the armored units were small, they represented a significant proportion of each country's contribution. Their presence, moreover, showed a strong inclination on the part of these countries, particularly those in Asia, to use armored units anywhere as part of a combined arms team. A balanced combat force was their goal regardless of the nature of the terrain.

CHAPTER V

Three Enemy Offensives

By late 1967 it appeared that the continued pressure of American and other free world forces in offensive operations and the South Vietnamese armed forces in pacification would eventually prevail. Enemy documents captured later indicated that the enemy had reached approximately the same conclusion and was so concerned about the outcome of his operations that he had decided to begin a general offensive to decide quickly the fate of South Vietnam. The new campaign was designed to inflict heavy losses on free world units and to destroy the South Vietnamese Army. The first objective was to launch a massive attack on all urban population centers and to seize and hold as many of them as possible. The enemy expected this offensive to be followed by a popular uprising that would reinforce Viet Cong and North Vietnamese units with deserting South Vietnamese troops.

Enemy Buildup

As early as the fall of 1967, intelligence sources had reported an increase in enemy road construction along the infiltration routes that crossed the border. In late fall the free world forces captured a document that described the enemy plan in general terms, but did not specify the timing. Large numbers of enemy troops and quantities of supplies were meanwhile being slipped into Saigon, Hue, and other cities as the South Vietnamese prepared to celebrate the lunar New Year, *Tet*, on 31 January 1968, welcoming the Year of the Monkey. *Tet* is both a solemn and joyous occasion; it is observed each year with exploding fireworks, feasts of rice cakes and other delicacies, and traditional visits to home villages. Enemy soldiers mingled easily with the crowds of holiday travelers and, aided by a well-organized network of agents, entered the cities.

There were other signs of an impending enemy offensive: captured members of the Viet Cong stated that the country would be "liberated" by *Tet*, and there was a sharp drop in the number of enemy deserters. Nearly forty attacks took place against outposts and towns in the upper Mekong Delta, where the Viet Cong often tested new tactics. By 24 January General Westmoreland, certain that the enemy would violate any *Tet* truce, had persuaded the U.S. and South Vietnamese governments to cancel a truce in the I Corps

Tactical Zone. Lt. Gen. Frederick C. Weyand, now commander of the II Field Force, Vietnam, was also concerned because units along the Cambodian border in the III Corps Tactical Zone were not having as many encounters with the enemy as they usually did. With General Westmoreland's approval, he shifted American forces back toward populated areas to improve close-in protection. By the eve of *Tet*, twenty-seven U.S. maneuver battalions were inside the so-called Saigon circle, within a 45-kilometer radius of Saigon; twenty-two maneuver battalions remained outside that radius.[1]

On 29 January cancellation of the truce in the I Corps Tactical Zone was announced. That night many free world bases in the I and II Corps zones were attacked. The next day the *Tet* truce was officially canceled throughout the country, and commanders were ordered to place troops on maximum alert and to resume offensive operations.

Before *Tet* armored units in all four corps tactical zones were in position to counter possible enemy attacks. American armored units had been shifted away from cities and were in position to block infiltration routes and quickly meet any major enemy attack. Vietnamese armored forces were stationed in or near cities to provide protection against anticipated attacks on population centers. Although on alert, the troops in the field expected nothing more than the usual attacks by indirect fire and occasional ambushes or probes. But the one flaw in the preparations was that no one, not even the staff at the Military Assistance Command, expected an attack of the magnitude and scope that developed in the first enemy offensive of 1968. At high command levels in Vietnam the attack had all the impact of the German Ardennes offensive of World War II.

For South Vietnamese and free world armored forces the battles of 1968 marked the acceptance of armor as an asset to the fighting forces in Vietnam. That acceptance was won on the battlefield by a demonstration of mobility and firepower that silenced all critics. When the enemy forced free world forces to move rapidly from one battle area to another, it was the armored forces that covered the ground quickly and in many cases averted disaster. Rapid movement was imperative in the early stages of the enemy attack, and the armored units were the first ground forces to reach the battlefield in almost every major engagement, although the winning of the battles eventually involved all forces. The story of these battles

---

[1] One of the latter units, the 2d Squadron, 17th Cavalry, arrived in mid-December 1967 and was immediately pressed into action at Song Be in the northern III Corps Zone. The squadron, equipped with wheeled vehicles as part of the 101st Airborne Division, fought with distinction during *Tet*.

points to the value of armored units as reaction forces. There was no set plan on the free world side in the early stages of the enemy attack: to defend was the purpose and armored forces made a substantial contribution to that defense. The three battles of the *Tet* offensive of 1968 described here serve as illustration of all the others in which armored units fought as the major counterattack force. One was a U.S. cavalry action, another a U.S. mechanized infantry battle, and the last a South Vietnamese cavalry attack. All three demonstrate the value of mobility and armor-protected firepower when they are aggressively used.

First Offensive: Tet 1968

The early attacks in I and II Corps Tactical Zones on 29 January signaled to free world forces that the enemy offensive had begun. In northern I Corps, where Khe Sanh was beleaguered, the few available armored elements were used inside the base to reinforce weak points in the perimeter and to repulse enemy ground attacks. During the relief of Khe Sanh, air cavalry units played a major role in finding the enemy, adjusting supporting fire, and keeping continuous pressure on the enemy. *(Map 11)* Air cavalry reconnaissance was so successful that no helicopters were lost during the airmobile assaults of the relief forces.

Farther south, the South Vietnamese armored forces broke up an attack on Quang Ngai City; by acting quickly with overwhelming firepower they drove the enemy from the city in less than eight hours. In a two-day battle in what was known as the Pineapple Forest near Tam Ky later in February, ground and air cavalry units killed some 180 of the enemy and had only one cavalryman wounded. Air cavalry elements found and fixed the enemy forces while ground cavalry attacked and destroyed them.

In Hue, the ancient capital of Vietnam, air cavalry, ground cavalry, U.S. Marine armor, and South Vietnamese armor participated in the longest battle of *Tet*, which lasted twenty-six days. The fighting, confined to a thickly populated area, was sustained and intense. Tank crews were so shaken by multiple hits from rocket propelled grenades—as many as fifteen on some tanks—that crews were changed at least once a day. Armored units were in constant demand and often expended their vehicle ammunition loads in a few hours. Their support of the infantry and marines, who cleared the city, provided the firepower advantage.

While battles were raging in I Corps Tactical Zone the highlands and coastal plain in II Corps were also under heavy attack.

Khe Sanh
Hue
Pineapple Forest
Tam Ky
Quang Ngai
Kontum
Pleiku
Ban Me Thuot
Phuoc Binh
Ben Cat
Bien Hoa/ Long Binh
An My
SAIGON
Phan Thiet
My Tho
Vinh Long
Phu Vinh
MAJOR ARMOR BATTLES
TET, 1968
Battle
0
100 MILES
0
100 KILOMETERS

MAP 11

Pleiku withstood Viet Cong assaults for five days, thanks primarily to the combined arms teams defending the city. Tankers, cavalrymen, artillerymen, and engineers from American units fought side by side with South Vietnamese cavalry and infantry. Near the border between II and III Corps, South Vietnamese cavalrymen traveled 100 kilometers through ambushes in eleven hours to the seacoast city of Phan Thiet, where they battled for eight days. South Vietnamese cavalrymen fought in Ban Me Thuot, and there again their firepower and aggressiveness was a deciding factor. The battle for the provincial capital of Kontum in II Corps involved both American and South Vietnamese cavalry and tanks. Before this battle, Troop A, 2d Squadron, 1st Cavalry, had planned and rehearsed, down to squad level, various strategies for the defense of Kontum; several of these were swiftly put into effect shortly after the attacks began.

The one-day warning provided by attacks in the north were barely enough time to alert units in the III Corps Tactical Zone when the full fury of the offensive erupted in the Saigon area. Armored units participated in all major battles in III Corps, and in thirty-seven out of the seventy-nine separate engagements. At Thu Duc, a Saigon suburb, a cavalry task force of students and faculty from the South Vietnamese Armor School defeated the enemy in bitter street fighting. On 1 February U.S. cavalry and infantry met a large enemy force at An My, north of Saigon. In a two-day battle this combined arms team mauled elements of a Viet Cong regiment. In the battle of Ben Cat, U.S. cavalry platoons converged on the town from two directions during a night attack and drove the enemy away. Mechanized infantrymen cleared the enemy from the racetrack in the center of Saigon, and in an afternoon and evening battle in a cemetery, fighting hand-to-hand among the tombstones, they killed 120 of the enemy. It was in the suburbs of Saigon, however, that the fighting in III Corps reached a climax; it was there, in the critical approaches, that cavalry and mechanized infantry decided the fate of the city.

Battle of Tan Son Nhut

About 2100 on 30 January, Lieutenant Colonel Glenn K. Otis, commander of the 3d Squadron, 4th Cavalry, was ordered to send one troop to aid Tan Son Nhut Air Base on the north side of Saigon. *(Map 12)* Colonel Otis directed the Troop C commander, Captain Leo B. Virant, to prepare for this mission and to put his troops on short notice alert. Troop C was at Cu Chi, twenty-five kilometers northwest of Saigon, with one of its platoons securing the Hoc Mon bridge, ten kilometers closer to Saigon on National Highway 1.

At 0415 Colonel Otis was ordered to commit Troop C to block a Viet Cong regiment that had attacked Tan Son Nhut Air Base. The troop was on its way in fifteen minutes and while it was en route was ordered to destroy enemy forces attacking the air base itself. Troop C was to be under the operational control of the Tan Son Nhut commander: a guide from that headquarters was to meet the unit at the south side of the Hoc Mon bridge and lead it to the air base. To avoid ambushes along Highway QL-1, Colonel Otis flew over the unit and, dropping flares to discourage the enemy, guided it cross-country to the Hoc Mon bridge. Troop C then passed through its own 1st Platoon, which remained at the bridge, met the guide from Tan Son Nhut, and moved toward the air base. Colonel Otis returned to Cu Chi to refuel and rearm.

The only information provided to Troop C was that a large enemy force had attacked from west to east across Highway 1 and penetrated the airfield defenses. As the troop approached the air base about 0600, it came under heavy small arms, automatic weapons, and rocket grenade fire. The cavalrymen attacked to split the enemy force where Highway 1 passed the southwestern gate of Tan Son Nhut. This move isolated approximately 100 Viet Cong inside the air base and kept the main body outside the gate. The full brunt of the enemy attack now fell on Troop C. In the first few minutes several tracked vehicles were hit and troop casualties were heavy. Captain Virant was seriously wounded in the head. The remaining elements of Troop C kept firing and succeeded in slowing the assault, but enemy fire forced some members of the troop out of their damaged vehicles into a ditch alongside Highway 1. Unable to communicate with the air base, the troop called the squadron at Cu Chi for help.

Colonel Otis was present when the call came in and immediately started back to Tan Son Nhut, calling 25th Infantry Division headquarters for release of Troop B from its mission of guarding the Trang Bang bridge, fifteen kilometers northwest of Cu Chi. Instead, the 1st Platoon, Troop C, was released from the Hoc Mon bridge, and Colonel Otis directed it to move to Tan Son Nhut immediately. In addition, he ordered the squadron's air cavalry troop to support Troop C. Flying over the area, Colonel Otis observed Troop C deployed in an extended column formation along Highway 1, with four tanks and five personnel carriers burning. Even from the air, he could see that the unit was hard pressed.

Ammunition was running low in Troop C when the air cavalry

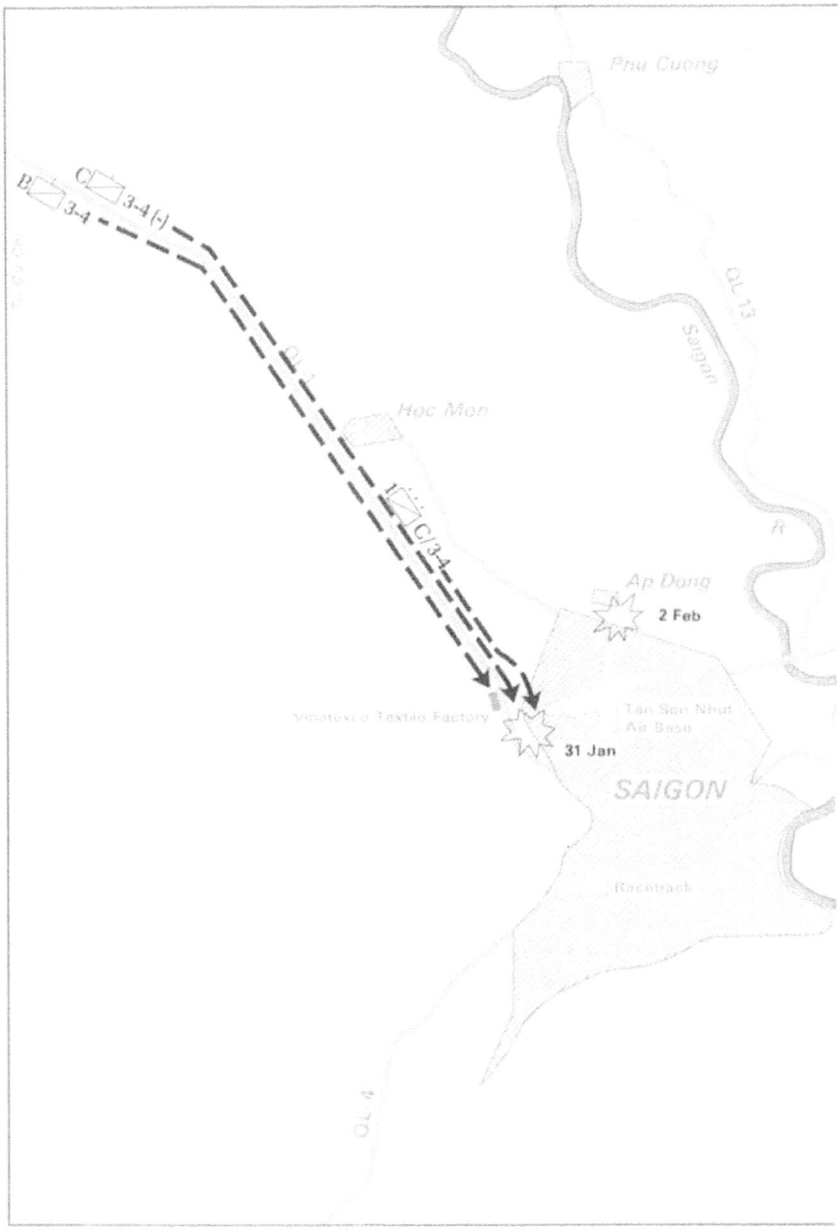
Phu Cuong
C 3-4 (-)
B 3-4
QL 13
Saigon
R
QL 1
Hoc Mon
C/3-4
Ap Dong
2 Feb
Tan Son Nhut
Air Base
31 Jan
SAIGON
Racetrack
QL 4

MAP 12

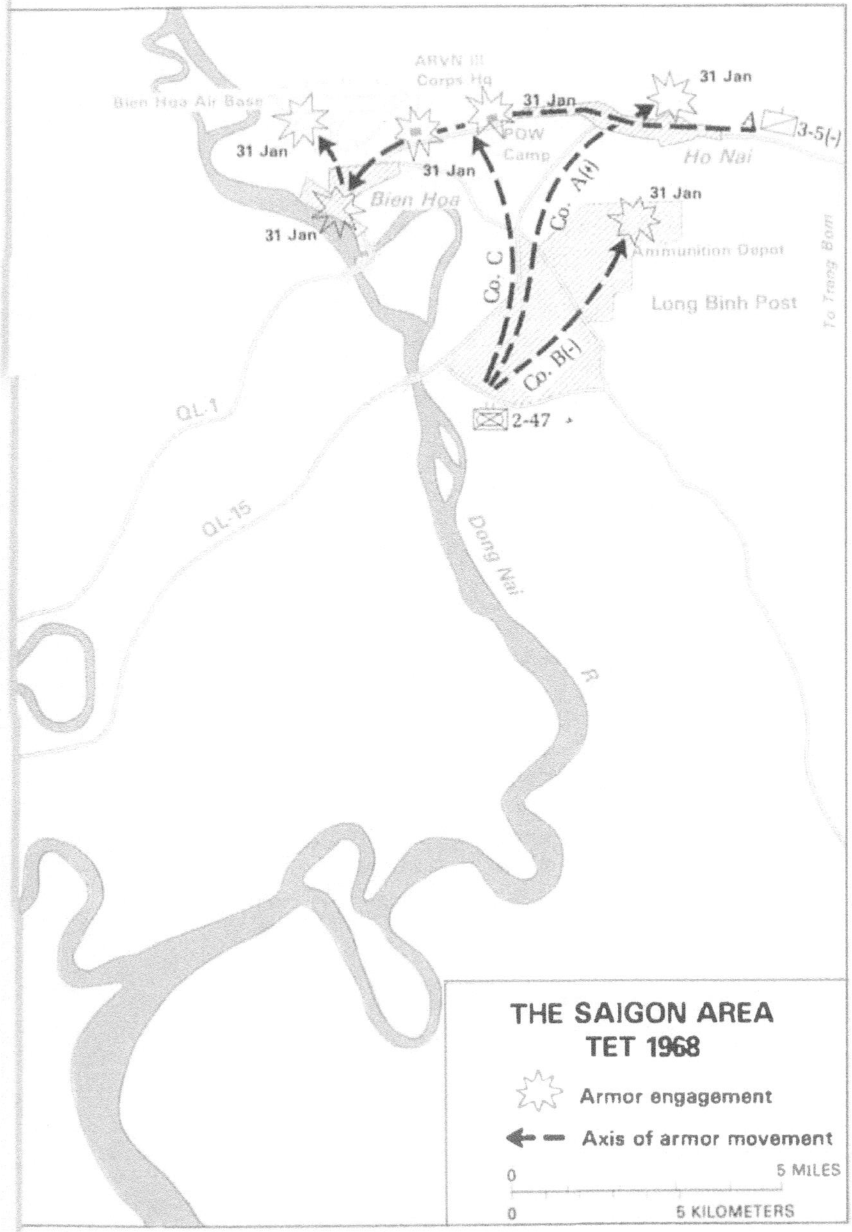

Bien Hoa Air Base
ARVN III Corps Hq
31 Jan
31 Jan
31 Jan
31 Jan
31 Jan
31 Jan
POW Camp
3-5(-)
Ho Nai
Bien Hoa
Co. C
Co. A(+)
Co. B(-)
Ammunition Depot
Long Binh Post
To Trang Bom
2-47
QL-1
QL-15
Dong Nai
R
THE SAIGON AREA
TET 1968
Armor engagement
Axis of armor movement
0
5 MILES
0
5 KILOMETERS

helicopters arrived. Gunships immediately attacked the enemy positions, firing rockets and machine guns. As the last few hand grenades were being thrown by the ground troops, two helicopters from Troop D, under heavy automatic weapons fire, landed with ammunition and took off the wounded. At 0715 the 1st Platoon of Troop C arrived. Colonel Otis directed it onto the airfield and then south to the left flank of Troop C, where it attacked west, relieving some of the pressure.

Colonel Otis ordered Troop B, commanded by Captain Malcolm Otis, to leave the Trang Bang bridge and move at top speed down Highway 1 to Tan Son Nhut, forty-seven kilometers distant. Captain Otis's troop, traveling fast, reached the battle area in forty-five minutes. When the troop arrived, it executed a column right toward the west at the Vinatexco Textile Factory, which put it parallel to the northern flank of the Viet Cong attack. With all vehicles of Troop B off the main road and strung out in column, Colonel Otis directed a left flank movement that brought them on line on the flank of what was later estimated to be at least 600 enemy soldiers. Troop B attacked with such intensity that many of the enemy immediately fled to escape the fire. Some attempted to reach a tree line three kilometers to the west across open rice paddies, but Captain Otis sent his 3d Platoon and Troop D gunships to cut them off. Caught in a cross fire between Troops B and C and heavy air and artillery fire, the Viet Cong were pinned in place.

The battle reached a climax at about 1000, when Troop B's flank attack began to take its toll of the enemy. Although fighting went on until 2200, from 1300 to 2200 the primary business was mopping up—hunting down the confused and beaten enemy. Subsequent sweeps of the battle area produced over 300 enemy dead, 24 prisoners, hundreds of enemy weapons of all kinds, and enough equipment and ammunition to fill a five-ton truck. At 1400 on 31 January, Colonel Otis was finally able to rendezvous with the Tan Son Nhut command. From the time Troop C called, asking for help, until 1400, the action was completely independent of the Tan Son Nhut command, controlled only by the cavalry squadron commander.

The movement of Troop C, 3d Squadron, 4th Cavalry, to Tan Son Nhut Air Base was a difficult night maneuver that achieved tactical surprise. When the enemy proved to be more formidable than had been expected, timely reinforcement, gunship and artillery support, and aerial resupply by units under command of the squadron turned the tide. The fact that the battle was fought by one unit and directed by one commander greatly facilitated control, but the deciding factor was the cavalry firepower that dominated the action.

Colonel Otis's cavalrymen were to have no rest after Tan Son Nhut, for the 3d Squadron, 4th Cavalry, was soon involved in its second major battle in as many days. This time it had help from Troop A, 1st Squadron, 4th Cavalry, which, after constant fighting throughout Saigon, was sent to Tan Son Nhut under operational control of the 3d Squadron, 4th Cavalry. Shortly after daybreak on 2 February, the 2d Battalion, 27th Infantry, the 3d Squadron, 4th Cavalry, and Troop A, 1st Squadron, 4th Cavalry, were ordered to conduct a reconnaissance in force through Ap Dong, a village north of Tan Son Nhut. Following tactical air strikes and an artillery preparation, the operation commenced shortly before noon when Troop A moved with the infantry to search the village. After a bitter all-day battle, the mounted units and the infantry succeeded in clearing the village and securing the northern perimeter of the air base.[2]

Battle of Long Binh and Bien Hoa Area

The cavalry battles on the northwestern side of Saigon blunted one of two major attacks toward the capital. The enemy made his other effort in the Long Binh–Bien Hoa area, twenty-two kilometers northeast of Saigon. This huge, sprawling, logistical and command complex had been carved out of a rubber plantation early in the war. It contained II Field Force headquarters; III Corps South Vietnamese Army headquarters; the U.S. Military Assistance Command, Vietnam, III Corps advisory headquarters; U.S. Army, Vietnam, headquarters; Bien Hoa Air Base; and the mammoth Long Binh Logistics Depot. *(Map 12)*

On the night of 30 January, the 2d Battalion, 47th Infantry (Mechanized), commanded by Lieutenant Colonel John B. Tower, moved to Long Binh as a reaction force for the 199th Light Infantry Brigade. The battalion dispersed by companies to screen for possible Viet Cong and North Vietnamese infiltrators. The battalion was also instructed to be prepared to reinforce critical command, control, and logistical installations in the giant complex.

At 0100 on 31 January, at the same time Tan Son Nhut was attacked, a well-coordinated assault was launched against the Long Binh–Bien Hoa complex from all points of the compass. At least four Viet Cong or North Vietnamese Army battalions were in-

[2] For the battles in Saigon during the *Tet* period, the 3d Squadron, 4th Cavalry, was awarded a U.S. Presidential Unit Citation. Four of its members received the Distinguished Service Cross, seven of them the Silver Star, and many the Bronze Star.

volved, and for a time the command post of the mechanized battalion was flooded with reports of enemy action. All elements were placed on alert and at 0445, the battalion was ordered to reinforce the Long Binh ammunition depot, the III Corps South Vietnamese Army headquarters, and the III Corps prisoner of war compound.

Company B, minus one rifle platoon, arrived at the ammunition dump at 0630 and, after coordinating with the defenders, moved into the area. The company was greeted by sniper fire, which knocked several infantrymen off their vehicles and diverted attention from the Viet Cong who were placing satchel charges in and around the ammunition bunkers. Despite the knowledge that they were operating in the midst of tons of explosives, men of the 2d Battalion, 47th Infantry (Mechanized), ran from bunker to bunker retrieving explosive charges, as other troops sought out and killed the snipers and sappers. At 0750 several bunkers in the storage area blew up and Company B suffered four casualties. The remainder of the day the company continued working through the area, clearing bunkers and forcing the enemy to withdraw. The II Field Force ordered Company B to remain in the ammunition dump as a security force, despite Colonel Tower's plea that they be used elsewhere.

Carrying out the battalion's second mission, Company C, 2d Battalion, 47th Infantry, arrived at 0554 at the III Corps South Vietnamese Army headquarters, which was under heavy enemy assault. Attacking from march column, Company C crashed into the flank of the Viet Cong forces, pinning them down with machine gun fire while the infantry dismounted and overran the position. The final assault required house-to-house fighting before the enemy was defeated. With the headquarters compound secure, at the cost of eight men wounded and one APC lost, the company moved to the prisoner compound east of Bien Hoa City. Again meeting strong resistance and fighting from house to house, the dismounted troopers finally overran the enemy positions in the late afternoon with a savage attack supported by the mounted machine guns. At 1730 the Viet Cong withdrew and Company C returned to the South Vietnamese corps headquarters, where it maintained security throughout the night.

With two companies fighting the enemy in widely separated locations, the 2d Battalion, 47th Infantry, was given a third mission—to reinforce a unit of the 199th Light Infantry Brigade near the village of Ho Nai on National Highway 1. Company A and the battalion headquarters moved north to link up with the infantry. The enemy attack was seriously threatening installations to the

TANK AND M113 DURING ENEMY ATTACK ON BIEN HOA, *Tet* 1968

east of Bien Hoa, and to relieve this pressure the mechanized infantry of Company A joined elements of the 4th Battalion, 12th Infantry, and 2d Battalion, 3d Infantry, in an attack north of Highway 1. The final successful assault was preceded by gunship and artillery fire and aided by machine guns from the APC's that forced the enemy from hastily dug foxholes. When the smoke and dust settled, bodies of forty-two of the Viet Cong were found, along with assorted weapons and equipment. At dark Company A, 2d Battalion, 47th Infantry, moved to a night position near the post of Long Binh.

The second act in the battle for Long Binh and Bien Hoa had begun while the 2d Battalion, 47th Infantry, was fighting its widely separated battles. Early on the morning of 31 January, Troop A, 3d Squadron, 5th Cavalry, was acting as a security force for an artillery unit at Fire Support Base Apple, twenty-eight kilometers east of Bien Hoa on National Highway 1. At 0100 the troop, commanded by Captain Ralph B. Garretson, was alerted to move. At 0230 Captain Garretson was directed to leave a platoon at the fire base and move the remainder of his troop toward Bien Hoa; he was to receive further instructions while he was en route. Troop A moved immediately, leaving the 3d Platoon to furnish security for the artillery.

There were early indications of things to come as the troop entered the town of Trang Bom on Highway 1 and ran into a company-size ambush. The fight lasted only five minutes as the troop, still moving, concentrated its fire along the roadside and rode through the ambush. The troopers, receiving sporadic fire along the way, reached a concrete bridge eighteen kilometers east of Bien Hoa. After the first tank crossed the bridge, a thunderous explosion dropped the span into the stream. The ACAV's had no trouble fording the stream, but with the exception of the one that had already crossed the tanks had to be left at the bridge.

Now out of radio contact with both the squadron headquarters and Bien Hoa, the troop continued on to the city to find the square filled with two companies of Viet Cong and North Vietnamese soldiers. The 1st Platoon charged through the enemy not realizing who they were. By the time Captain Garretson arrived, however, the enemy force had dispersed and opened fire, disabling two ACAV's. The 2d Platoon pushed aside the disabled vehicles and entered the square in a hail of machine gun fire. After this brief action the small cavalry force consisted of one tank and eight ACAV's.

Continuing to move, the troop was joined by the squadron commander, Lieutenant Colonel Hugh J. Bartley, who directed Captain Garretson to Bien Hoa Air Base. As he flew over the troop, Colonel Bartley spotted an enemy ambush just outside the base, with several hundred Viet Cong and North Vietnamese troops in the ditches near the southeast entrance. The men quickly moved off the road some thirty to forty meters behind the ambush, firing as they went, and destroying the enemy force on one side of the road.

After entering the base, the cavalry force was attached to a battalion of the 101st Airborne Division, which was attempting to reduce the enemy forces on the base perimeter. The cavalrymen were split between two companies of infantry attacking a position at the southeast corner of the air base. The battle lasted most of the day, culminating in a breakout attempt by the enemy that was stopped short by the cavalry troop. In this fight the 2d Platoon lost two ACAV's; the one tank lost two crews and took nineteen hits from rocket grenades but was still operational. For the rest of the day the troop was the reaction force for the air base defense. The next morning Troop A returned to Fire Support Base Apple; it had suffered five killed and twenty-three wounded.

The final curtain was rung down on the Long Binh–Bien Hoa battle by the 11th Armored Cavalry Regiment on 1 February. Although the regiment was conducting operations in the thick jungles of War Zone C, on the morning of 31 January the entire unit pulled out, consolidated, and moved over 103 kilometers to Bien Hoa, all in eight hours. The regiment had completely circled the Long Binh–Bien Hoa complex by 2100 that night. The next day U.S. airborne infantry and the 3d Squadron, 11th Armored Cavalry Regiment, swept through Bien Hoa and cleared the town. For the rest of the month the 11th Armored Cavalry remained in the Long Binh–Bien Hoa and Saigon areas.

Battles in Vinh Long Province

The *Tet* battles in the IV Corps Tactical Zone were characterized by the same intensity as those in other tactical zones, but in most cases they involved South Vietnamese Army units. In the provincial capital of My Tho, South Vietnamese cavalry forces along with U.S. and South Vietnamese infantry engaged in a bloody three-day battle in which over 800 enemy troops were killed. Farther south Vietnamese Army cavalrymen were ordered on one hour's notice to fight their way along the fifteen kilometers from their base camp to the city of Phu Vinh (Tra Vinh). Using massive firepower and without the aid of supporting infantry, they destroyed the enemy resistance in the city in less than twenty-four hours.

The *Tet* offensive in Vinh Long Province started at 0300 on 31 January when the Viet Cong attacked Vinh Long City and the compounds of South Vietnamese cavalry units, ninety-five kilometers southwest of Saigon in the delta. The 3d Squadron, 2d Armored Cavalry Regiment, swiftly secured its own compound and dispatched elements to assist the headquarters of the 2d Armored Cavalry Regiment, which was under attack. The 3d Squadron forces crashed through the enemy force around the headquarters perimeter, then attacked and routed the Viet Cong.

The squadron was then ordered to assist in the defense of Vinh Long Airfield, while regimental headquarters troops were to provide security for the southwestern portion of the city perimeter. After clearing the outer perimeter of the airfield, the 3d Squadron was ordered back into the city where it cleared the route from the airfield to the tactical operations center in the middle of the town. When the cavalry started to clear Vinh Long's main street, heavy automatic weapon and rocket propelled grenade fire from buildings stalled the mounted attack. Finally, the 3d Squadron linked up with a company of the 43d Ranger Battalion of the South Vietnamese Army, but in the face of intense fire the Rangers retreated, leaving the 3d Squadron to continue the assault on its own.

A Regional Forces company was sent to fight with the squadron, but it too refused to advance. After a direct order by the cavalry commander, the Regional Forces company reluctantly followed the ACAV's, but the attack bogged down. As evening came, the company departed and the 3d Squadron pulled back to protect secured areas of the city.

Early the next morning the squadron again attempted to clear the main street, still without infantry support. The fighting seesawed back and forth all day through the rubble, but little progress was made. Every move toward the main street had to be fought from house to house. Even when the squadron was reinforced by fighting elements of the regimental headquarters, the most that could be achieved was security of the tactical operations center and the provincial administrative center, and even that was a constant battle.

During the night of 1 February, South Vietnamese soldiers from the 3d Battalion, 15th Infantry, arrived by boat, and the next morning the 3d Squadron, 2d Armored Cavalry Regiment, and the infantry began clearing the main street. For the fourth consecutive day there was heavy fighting. With U.S. helicopter gunship support, and after five full days of frustration, the combined forces charged the Viet Cong positions and broke through. The cavalry troopers and a Ranger company quickly cleared the western section of the city and the remaining Viet Cong fled.

In five days of intense fighting South Vietnamese Army cavalrymen had adapted themselves to urban combat: in the beginning they did not even have the aid of supporting infantry. The courage and tenacity of individuals and the ability of the unit to bring firepower to bear in critical areas kept the Viet Cong off balance and finally enabled the cavalrymen to seize the initiative from the attackers.

The magnitude of the enemy *Tet* offensive of 1968 surprised the free world forces. Despite intelligence reports to the contrary, the enemy was considered incapable of mounting a countrywide offensive and such an attack was not given serious military consideration. That the Viet Cong and North Vietnamese could attack Saigon or seize and hold Hue for twenty-six days was believed impossible. Those who throught that an attack was coming did not foresee a widespread offensive. When the attacks came, armored units were sent into almost every significant action in South Vietnam. Mounted troops became quick reaction forces, attacking in large cities and destroying enemy units or driving them away. Armored forces screened the borders, blocking enemy infiltration routes, denied the enemy access to his objectives, and intercepted units along routes of resupply and withdrawal. Route security and convoy escort were performed by mounted units to keep roads open and insure that convoys reached their destinations. In *Tet* 1968, armor paid its way.

Second Offensive

In the period immediately after *Tet*, free world forces struggled to sort out units in the resulting confusion. Many, particularly armored units, were widely spread in defensive missions far from their normal operational areas. Losses, particularly of equipment, had been heavy, and many units needed time to replace materiel and train new men. For a time there was a shortage of M48A3 tanks, made up for in part by issuing M48A1 tanks to selected units. The hull and guns of the tanks were the same, but the M48A1 used gasoline and its reduced range and susceptibility to fire did not make it very popular.

Intelligence analysts indicated that the enemy was also replacing troop losses and rebuilding supply caches preparatory to launching another offensive. Sweep operations disrupted the enemy time schedule for renewed attacks, but larger enemy units remained hidden in base areas throughout South Vietnam. Anticipation of renewed enemy activity, particularly around Saigon and the Capital Military District, resulted in many false warnings with units alerted for attacks that never came.

To relieve the pressure free world forces attacked, with South Vietnamese Army units now participating in strength. In early April Operation TOAN THANG I was launched to drive the enemy units away from Saigon. Highly decentralized, the operation was characterized by small daylight search missions and night ambushes involving seventy-nine battalions in III Corps Tactical Zone alone. Intelligence information gathered in these actions revealed the imminence of a second enemy offensive. The *Tet* attack had suffered from poor coordination among the enemy units because strict security measures were employed. This time the enemy widely disseminated his plans, enabling the free world forces during TOAN THANG I to learn of them quickly from captured documents and prisoners. By late April all invasion and infiltration routes into Saigon and other key cities were watched and heavily guarded.

This alertness, the large-scale allied operations after *Tet*, and heavy enemy losses during *Tet* weakened the second enemy offensive in early May, which U.S. forces nicknamed "Mini-*Tet*." The main enemy effort was made in the Saigon area, where most of the combat took place. The 1st Squadron, 4th Cavalry, got into action early when, on 5 May 1968, it discovered three North Vietnamese soldiers seventeen kilometers north of Saigon and triggered a two-

M41 of South Vietnamese Army Advances on Enemy Positions in Saigon, May 1968

day battle with a North Vietnamese Army battalion. After massive air, artillery, and air cavalry support and constant mounted attacks by ACAV's and tanks, the combined arms team routed the enemy. By dark on the 6th, the North Vietnamese had gone—leaving over 400 dead on the battlefield.

Closer to Saigon, the 5th Battalion, 60th Infantry (Mechanized), and the 2d Battalion, 47th Infantry (Mechanized), again closed with the enemy in house-to-house fighting near the "Y" bridge on the west edge of the city.[3] The mechanized troops used every means of fire support available in six days of intense fighting that, curiously, was broken off each night by the enemy. The final major battle was fought on 10 May by the 5th Battalion, 60th Infantry, and the enemy was driven away from the city for the last time. Mop-up actions continued for a few days, but the battle for Saigon was over.

---

[3] Shortly before this fight, three tanks and an APC from the South Vietnamese 1st Squadron, 5th Cavalry, had engaged heavy enemy forces at the "Y" bridge. They were unable to convince the higher commands that they were in a fight. When one entire lane of the four lane bridge blew up, mechanized reinforcements were finally sent.

After the May offensive failed, the enemy retreated to base camps with the free world forces in pursuit. Another minor attack was made against Saigon on 25 May, but the attackers were quickly routed. Viet Cong and North Vietnamese troops emerged from hiding places to surrender. On 18 June the largest number of enemy troops surrendered—141 enemy soldiers turned themselves over to the South Vietnamese Army forces northeast of Saigon. Scattered and diminished fighting continued till late June. While the fighting was still going on, peace talks had begun in Paris on 13 May.

The May attacks were but a shadow of the *Tet* offensive in February, and had no apparent military objective. It appeared that small groups of enemy soldiers were dispersed over wide areas in an attempt to make their attacks appear heavier than they actually were. A number of rocket attacks were also employed against the capital to create an image of Saigon under seige. The principal target was Saigon, with attacks in other areas of the country designed to divert free world forces. It appears probable that the attacks were intended to influence the peace talks rather than to achieve a military goal.

Third Offensive

After the second enemy offensive, just as after the first, both sides stepped back to assess the result and repair the damages to men and equipment. For the free world forces this interim was unlike the previous one because an air of confidence and optimism prevailed. The recuperative means of the free world forces were greater than those of the enemy. No longer content to sit and wait for the Viet Cong and North Vietnamese to attack the populated areas, units went into the bush with a vengeance. Base areas were penetrated repeatedly. Thus the third and last enemy offensive during 1968 fell not on the population or on critical areas but directly on the free world combat units, who were well prepared.[4]

Recognizing the change in the free world military position, the enemy chose objectives that differed in two ways from those of earlier offenses. First, the offensive had limited goals. Second, the attacks were directed against U.S. troops, base camps, and equipment rather than at Vietnamese population centers. Intelligence after the attack revealed that a prime objective was to inflict heavy

[4] One of the most welcome improvements made during this period in U.S. units was the conversion of the M113 to the diesel-powered M113-A1. A main drawback of the M113, the danger of fire, was now reduced. Other free world units were scheduled for conversion later.

losses on U.S. troops in hope of winning a political and psychological victory.

The most sustained fighting of this third offensive took place in the III Corps Tactical Zone in Tay Ninh Province. By August 16,000 soldiers of the 5th and 9th Viet Cong and North Vietnamese Divisions had infiltrated into War Zone C, prepared to attack. The 1st Brigade, 25th Infantry Division, the largest free world force in the area, included the 1st Battalion, 5th Infantry (Mechanized); 4th Battalion, 23d Infantry (Mechanized); 3d Battalion, 22d Infantry; 2d Battalion, 34th Armor (–); and the 2d Battalion, 27th Infantry. Relying on the mobility of its armored units, the brigade was organized into combined arms task forces with attached tank, infantry, and mechanized infantry units heavily supported by artillery. From the fire bases around Tay Ninh City task forces sent out wide-ranging patrols to prevent or break up enemy movement.

For ten hectic days and nights beginning 17 August 1968, the 1st Brigade fought the enemy over an area of 1,500 square kilometers in the unfavorable summer wet weather. The fight started when the enemy made a night attack on a fire support base six kilometers north of Tay Ninh, using human wave tactics; the attackers were stopped short of the defensive wire by the overwhelming firepower of tanks, aircraft, artillery, and infantry. By daylight a heavy attack on Tay Ninh City was under way, but at the end of the day the enemy had been beaten back by two armored task forces and a cavalry troop. The enemy then shifted thirty kilometers to the east into some rubber plantations in the hope of bypassing the brigade and striking farther south. But the mobile task forces followed, and for nine days fighting ranged through the rows of rubber trees. Counterattack followed attack with such regularity that it became difficult to tell which was which. The only certainty was that the brigade was keeping the pressure on. Movement and firepower were the keys, aided in strong measure by the unity of command within the brigade.

In almost every fight, and they took place every day and every night, the enemy was severely beaten. There were exceptions. On 21 August a mechanized infantry company took on two North Vietnamese battalions and for more than an hour held its own. After suffering heavy losses and with one officer left alive, the company was forced to withdraw under cover of supporting artillery. Again, on 25 August a logistical convoy of eighty-eight vehicles was ambushed south of Tay Ninh. An armored force moved in and severely punished the enemy, but not before the convoy had lost many men and vehicles. In these actions the Viet Cong and North Vietnamese prevailed by weight of numbers.

M113 IN ACTION AT BEN CUI RUBBER PLANTATION, AUGUST 1968. *Hollow steel planking on sides of vehicle offer protection against shaped charges.*

By 27 August the third enemy offensive in War Zone C had ended in defeat for enemy units. It was largely the mobility and firepower of the 1st Brigade combined arms teams that made victory possible for the smaller American forces. The ability of U.S. forces to maneuver and mass rapidly to defeat a strong but slower enemy was the critical factor. The attacks were the last in the III Corps Tactical Zone. For the remainder of the year and well into the next the enemy stayed in Cambodia refitting badly beaten units.

The third enemy offensive, however, was not limited to the III Corps area. In I Corps far to the north the American 1st Squadron, 1st Cavalry, 23d Infantry Division, blocked the path of three regiments of the 2d North Vietnamese Army Division that intended to attack Tam Ky City. The squadron had previously fought insignificant battles west of the city, and consequently was maintaining protective security around Tam Ky by means of daily reconnaissance in force. It was a platoon in one of these sweeps, operating with a troop of South Vietnamese cavalry, that started the fighting in one of the hardest battles the 1st Squadron, 1st Cavalry fought against the North Vietnamese.

On the morning of 24 August 1968, Lieutenant Thomas Guiz,

leader of 2d Platoon, Troop A, 1st Squadron, 1st Cavalry, was operating west of Tam Ky with a South Vietnamese troop from the 3d Squadron, 4th Cavalry. At 0925 the South Vietnamese cavalry received intense fire from a clump of trees on high ground surrounded by rice paddies. In the first few seconds, one APC was hit several times by recoilless rifle fire and exploded in flames. The Vietnamese cavalry, calling for reinforcement from the 2d Platoon, fired into the enemy position.

Lieutenant Guiz and his platoon responded, but so heavy was the enemy fire that it took two air strikes and air cavalry support before the platoon could maneuver abreast of the Vietnamese cavalry troop. When the air strikes lifted, the combined forces then assaulted the hill in a line formation. Several vehicles were hit by recoilless rifle fire and the attack stalled. The enemy force was too strong to be pushed off the hill by frontal assault so the cavalrymen withdrew and awaited reinforcements from the 1st Squadron, 1st Cavalry. Lieutenant Colonel Raymond D. Lawrence, commanding the 1st Squadron, ordered Troop C and the remainder of Troop A into the battle. The enlarged cavalry forces probed the area with automatic weapons fire and tried to encircle the high ground, but were driven away by heavy fire from antitank weapons. Unable to close with the enemy, the cavalrymen dismounted and continued the attack on foot, but even after savage bunker-to-bunker fighting little ground was gained.

Throughout the afternoon Colonel Lawrence had repeatedly requested a company of infantry to assist his cavalrymen in the fight. At 1630 an infantry company became available to make a combat assault if a secure landing zone could be provided. A landing zone 2,000 meters away was selected, but when Troop C moved toward it, the men ran into the heaviest fighting of the day. "We had to cross a small forest to get to the LZ," stated Private First Class Mark Bellis, "and as soon as we were in the middle, we were right up ambush alley." The attack came from all sides and the cavalrymen formed their vehicles in a circle to fire in all directions. Again and again, the North Vietnamese infantry rushed the circle only to be stopped by the heavy fire. As darkness approached, Colonel Lawrence realized that the landing zone could not be secured, and the airmobile infantry, now two companies, were therefore sent to Hawk Hill, base camp of the 1st Squadron, 1st Cavalry. Both companies readied themselves for anticipated fighting in the morning.

Troop C fought its way out of the small forest area back to its original position, and there joined Troop A after dark. Although supplies were now critically needed by all units, the policy of the

23d Infantry Division did not permit resupply by CH–47 helicopter if hostile fire was being received by a ground unit. The cavalrymen, therefore, were forced to break away from the enemy to secure a landing zone. The two troops then set up a single position near the battlefield to await resupply. The South Vietnamese 3d Squadron, 4th Cavalry, withdrew to Tam Ky to defend the city. Fighting on 24 August had been fierce, with over 200 North Vietnamese killed, but the enemy was still present and in strength.

Colonel Lawrence's plan was to place an additional cavalry troop, Troop B, in a blocking position, and to attack with the other two cavalry troops, reinforced with infantry, which would arrive at first light. At 0615 while en route to the blocking position, Troop B came under attack from a North Vietnamese force. The cavalrymen, quickly gaining fire superiority, left one platoon to deal with the enemy as the other two platoons drove on. Reaching the proposed blocking position, the cavalrymen found themselves in the middle of a North Vietnamese regimental headquarters. They drove their vehicles through the position, firing in every direction, but enemy fire was so intense that they were finally forced to regroup and form a circle. As on the previous day, the North Vietnamese tried to overrun the vehicles but were repulsed. Although surrounded, the troop had disrupted the enemy command and control facilities.

While Troop B's battle raged, Troops A and C started to advance toward their assigned positions. Almost as soon as it left the night camp, Troop A came under heavy rocket grenade and recoilless rifle fire. Ground gained in this sector was measured in meters, as infantry and cavalry unsuccessfully battled the enemy for the high ground. Troop C, which met little resistance at first, was stalled when its attached infantry was pinned down by heavy small arms fire. Thick undergrowth and rough terrain slowed the cavalry attempt to help the infantry, but after an all-day battle both cavalry and infantry managed to pull back. With darkness approaching, once again the resupply needs of the armored force and the division policy regarding CH–47 helicopters forced the cavalry to disengage. In positions near the battlefield, the U.S. forces resupplied and readied themselves for more fighting. The day had been expensive for the North Vietnamese, who lost 250 killed.

The plan for 26 August called for the entire force to shift southwest, then attack north with all three cavalry troops and the attached infantry. If the North Vietnamese were still in their trenches, the flanking attack was designed to surprise them. Following a twenty-minute artillery preparation, the squadron and its infantry attachments attacked the enemy right flank; resistance was light—the enemy had left a few rear guard troops and withdrawn from the battlefield. During the daylong scattered fighting that followed, Team B captured two North Vietnamese soldiers who related that over 400 of their troops had fled to the northwest the previous night. Logistical needs once again outweighed tactical needs, and the squadron halted to resupply.

Colonel Lawrence sent three teams to the northwest on 27 August. Troop F (Air), 8th Cavalry, screening to the front, reported the presence of a North Vietnamese antiaircraft unit. A prisoner later related that the North Vietnamese unit had one rocket propelled grenade crew attached, and that the opening round was to be the signal for everyone to open fire. Team A, deployed on line, laid down a base of fire, and advanced on the enemy at full speed. The rocket grenadier took aim at the lead American vehicle and pulled the trigger—misfire! With no signal given, the North Vietnamese unit panicked. "We just drove right in," stated Lieutenant Guiz. "We captured two caliber .51 antiaircraft machine guns and found a lot of enemy packs. The team killed or captured all but three men in the unit."

The enemy force, estimated to number around 1,300 men before the start of the battle, had had enough, and in the darkness that followed the North Vietnamese melted back into the hills. The successful cavalry sweep operations had intercepted the enemy movement toward Tam Ky, and forced a fight before the North Vietnamese were able to reach their objectives. Armored firepower and air and artillery support had inflicted crippling losses and had stopped this part of the enemy third offensive as it began.

Aftermath

In the wake of these enemy offensives, the free world forces found themselves in control of the battlefields. For armored units the offensives, particularly *Tet*, had brought a demonstration of everything that armor proponents had claimed. The ability of armored units to move fast with overwhelming firepower had been of the utmost importance throughout the offensive.

Armored units in combined arms task forces would now spend most of their time in the bush pursuing the weakened Viet Cong and North Vietnamese. A new American armored unit, a mechanized brigade, was being sent to Vietnam. The South Vietnamese armored force was forming seven more cavalry regiments.[5] Even

[5]The first of these new regiments, the 11th Armored Cavalry, was deployed on 18 July 1968. Out of respect to superstition there was no 13th regiment. The South Vietnamese armored force also began a change to five APC's per troop (U.S. platoon)

more important, in the redeployment planning that was soon to begin, armored units, both air and ground, were scheduled as last to leave Vietnam because they provided mobility and firepower at far less cost in manpower than any other type of unit. It had taken a long time to convince the Army, but there was no longer any doubt about the utility of armored forces in Vietnam's counter-insurgency and jungle warfare. The three offensives brought about a real change in the acceptance of armored units and ended the longambivalance toward armor in Vietnam.

The campaigns of 1968—*Tet,* and the second and third enemy offensives—were the last concerted enemy offensives against the combined free world forces in Vietnam. In a period of seven months, the Viet Cong and North Vietnamese forces were severely defeated; over 60,000 enemy soldiers lost their lives without making any tangible gains on the ground. The free world forces struck back both during and after this campaign with a strength that carried the battle to the borders of South Vietnam and beyond. The soldiers of the Republic of Vietnam gained new confidence in the very battles that were designed to persuade them to desert to the enemy.[6]

a move which gave a squadron (troop) twenty-two tracked vehicles and significantly increased firepower.

6 But they also had to discontinue pacification support missions until late in 1968.

CHAPTER VI

The Fight for the Borders

Changing Strategy

In the aftermath of the 1968 enemy offensives both sides changed strategy. For the North Vietnamese, the change was a necessity brought on by heavy losses in men, equipment, and supplies. With the Viet Cong underground organization exposed or destroyed in many areas, main force units short of men, equipment, and leaders, and the logistical system drained of supplies, the enemy had no choice but to retire to sanctuaries. On the free world side, U.S. troop strength, despite some losses, reached a new high of nearly 550,000 in April 1969, and the South Vietnamese Army gained confidence from new weapons and freedom of action. It was the time, therefore, to pursue the enemy into his sanctuaries and keep up an unrelieved pressure that would prevent his returning.

The new strategy of the free world forces was applied under the leadership of General Creighton W. Abrams, who succeeded General Westmoreland as Commander, U.S. Military Assistance Command, Vietnam, in the summer of 1968. By background and training General Abrams was the man for the job. One of the great commanders of small armored units in World War II, he subsequently commanded an armored division and a corps, and served as Vice Chief of Staff of the U.S. Army. For a year before he assumed command he was General Westmoreland's deputy, and concentrated his attention on the Vietnamese Army. His rapport with South Vietnamese leaders was excellent, his confidence in the South Vietnamese Army was a great boost to Vietnamese morale, and his conviction that the South Vietnamese were capable of a much broader participation in the war than had been allowed them in the past augured well for the new strategy.

When the enemy forces fled from the battlefields, they regrouped in base areas inside South Vietnam. With virtually no large Viet Cong and North Vietnamese units in the field, free world forces began to engage in continual small unit actions to locate and defeat the remaining enemy forces. Increased American strength and the growing strength of the South Vietnamese Army also allowed free world forces to penetrate and destroy the enemy base system in South Vietnam. As the tempo of these operations increased, the enemy fled once again—this time to sanctuaries in Laos and Cambodia where refitting, resupplying, and training could be carried out without interference.

By mid-1969 the enemy was operating from these sanctuaries, venturing occasionally into South Vietnam as logistics permitted. These forays, known as high points, were preceded by enemy logistical buildups of enough supplies to support the high points. The enemy tactic of sticking out a so-called logistics nose, followed by troops who were supported by the buildup, became a familiar one. In 1969 and 1970 the free world countertactic was to cut off the logistics nose when possible and thus frustrate the enemy attack that was intended to follow.[1]

American and South Vietnamese armored forces in the air and on the ground played an important role in the fight to prevent a logistical buildup and to seal the borders of South Vietnam. Their mobility and heavy firepower enabled them to operate in small groups with less chance that any single unit would be overwhelmed. This mobility also enabled them to disperse over wide areas, yet mass quickly when the enemy struck.

Armored Forces Along the Demilitarized Zone

In mid-1968 the last major U.S. tactical unit that was to be sent to Vietnam, the 1st Brigade, 5th Infantry Division (Mechanized), was arriving from Fort Carson. By 1 August 1968 the five combat units of the brigade were in Vietnam: the 1st Battalion, 61st Infantry (Mechanized); the 1st Battalion, 77th Armor; the 1st Battalion, 11th Infantry; the 5th Battalion, 4th Artillery (155-mm., self-propelled); and Troop A, 4th Squadron, 12th Cavalry. Within a few days they started combat operations in the I Corps Tactical Zone, immediately south of the Demilitarized Zone between North and South Vietnam.

The North Vietnamese Army units along the Demilitarized Zone were unaccustomed to fighting U.S. armored forces. The U.S. Marines operated in this region, but used their armor in small groups whose primary role was support of dismounted infantry. Thus, the 1st Brigade enjoyed some immediate success against enemy troops who tried to stand and fight, believing themselves to be facing tank-supported infantry as before. These encounters were usually one-sided, with the North Vietnamese losing significantly in men and equipment. However, the enemy quickly learned the futility of a stand-up fight against the consolidated mechanized

1 General Abrams was the first to recognize officially the enemy tactic and to define the countertactic to cut off what he termed "the enemy's logistics nose."

force of the brigade, and changed his tactics to avoid the mounted formations.

Emphasizing offensive action away from fixed bases, the 1st Brigade attacked the enemy whenever and wherever he could be found. After only two months of combat the brigade received a letter of congratulations from General Westmoreland, newly appointed Army Chief of Staff, who wrote that the unit's actions had "demonstrated the Brigade's readiness to take its place with other veteran units in Vietnam."

By late 1968 1st Brigade operations in the I Corps area were fairly illustrative of small unit actions throughout the country. The brigade concentrated on rooting out the Viet Cong underground organization and breaking up guerrilla units by sending battalions into areas whose limits corresponded with local Vietnamese district boundaries. Liaison officers from the American battalions were assigned to each district and direct communications were established between battalion and district. A close association was thus developed between Vietnamese Regional and Popular Forces and U.S. battalions. Perhaps most important of all, the brigade began to develop a network of village agents that was to prove invaluable in providing timely information about the Viet Cong organization.

Mechanized infantry was particularly successful in ferreting out the Viet Cong. Mechanized companies could move rapidly through search areas and quickly cordon off a village suspected of harboring Viet Cong. Local Vietnamese forces operating with the American troops usually conducted the detailed search. When it was over the mechanized company would move on, sometimes many miles, to place a cordon around another village.

As these operations became more successful local government improved and Viet Cong village organizations collapsed. Members of the Viet Cong could no longer safely submerge in village populations, and when they fled to the countryside they were hounded by American infantry and artillery. Although it was certainly not the efforts of the mounted infantry alone that drove the Viet Cong from the villages, the mechanized troopers performed a task that neither police, local militia, nor standard U.S. infantry could accomplish alone. One Vietnamese district chief said: "On a good day, the U.S. mechanized infantry may not always get here quite as fast as airmobile infantry—but they stay with us longer and with more firepower."

After the early encounters, combat in the 1st Brigade's area of operation was light. Since lack of vegetation made the M113 visible for hundreds of yards, particularly on a moonlit night, it was a

PREPARING NIGHT DEFENSIVE POSITIONS ALONG THE DEMILITARIZED ZONE. *Men of 1st Battalion, 61st Infantry, dig foxholes but vehicles are left in the open to allow maneuver.*

simple matter for the enemy to bypass the vehicles. The brigade countered this tactic by saturating an area with four-man patrols. Each mechanized infantry company was required to have a minimum of twenty ambush patrols of four men each night. Commanders briefed their troops at the noon meal; patrols then mounted armored personnel carriers and were taken on a ground reconnaissance of each position. At dusk, while they were still visible to enemy in the area, the M113's were again dispatched on designated reconnaissance routes. Immediately after dark, while the APC's were moving, each four-man patrol dismounted and established its ambush position. This technique made it difficult for an observing enemy to detect ambush positions because the vehicles never stopped moving during the reconnaissance.

After the patrols were in place, the M113's formed platoon night defensive positions and prepared to move to the assistance of any patrol that ambushed an enemy force. Because ambush patrols were close together, usually separated by a rice paddy or a dike, the enemy could not bypass all of them. Establishing four-man patrols was a calculated risk since they had little staying power, but once the patrols had engaged the enemy, reinforcements moved according to plan. Upon hearing the first round fired, the vehicles nearest the ambush took the most direct route to the fight, with headlights blazing. Later, a tabulation of all fights during the use of this technique showed that the longest time lapse, from the first round fired to the arrival of the armored cavalry assault vehicles, was less than four minutes. It was the speed of the M113 that permitted the American forces to take the risk of setting up four-man patrols.

SHERIDAN M551 AND CREW MEMBERS OF THE 3D SQUADRON, 4TH CAVALRY

The Sheridan

In early 1969 the U.S. Army introduced a new combat vehicle into its armored forces in Vietnam—the General Sheridan M551. Cavalry commanders in Vietnam had long expressed a need for an amphibious tracked vehicle with more firepower than the armored cavalry assault vehicle but with the same mobility.[2] The Sheridan was a partial answer: it was to replace M48 tanks in cavalry platoons of divisional cavalry squadrons and ACAV's that had been substituted for tanks in cavalry platoons of regimental cavalry squadrons. The regimental squadrons of the 11th Armored Cavalry retained their M48A3 tanks in the squadron tank companies.

Designed as an antitank weapon for airborne forces, the Sheridan was sent to Vietnam without its primary antitank missile equipment aboard. Its armament consisted of a 152-mm. main gun, firing combustible-case ammunition with several different warheads, a .50-caliber M2 machine gun at the commander's station, and a 7.62-mm. machine gun mounted with the main gun. The Sheridan had a spotty development history, characterized by difficulties with

[2] Deployment of the Sheridans and installation of diesel engines in the M113's were not the only efforts to improve mobility. In May 1968 three air cushion vehicles were tested in the delta. Nicknamed the 39th Cavalry Platoon, the vehicles were so beset by maintenance problems that the test ended inconclusively

the complex electronics gear associated with its antitank missile system and problems with the combustible cases for its main gun rounds. The missile system was not a problem in Vietnam—it was not used—but the combustible case gave persistent trouble.

As early as 1966 the Army staff in Washington was pressing the U.S. Army, Vietnam, to accept the Sheridan for its cavalry units. At that time the U.S. commander in Vietnam demurred on the grounds that since no main gun ammunition was available the vehicle was no more than a $300,000 machine gun platform, not as powerful and agile as the M113. When main gun ammunition was finally available in 1968, however, plans to equip two divisional cavalry squadrons, the 1st and 3d Squadrons of the 4th Cavalry, were approved. Neither squadron wanted the Sheridan because it was suspected of being highly vulnerable to mines and rocket propelled grenades and could not break through jungle like the M48A3. General Abrams, during a visit to the 11th Armored Cavalry in late 1968, mentioned this fact to the regimental commander, Colonel George S. Patton. Colonel Patton suggested that the vehicle would receive a better test if the Sheridans went to a divisional squadron and a regimental squadron. General Abrams agreed, and sent the first Sheridans to the 3d Squadron, 4th Cavalry, and the 1st Squadron, 11th Armored Cavalry.

Both units started training in January 1969. The new vehicles were accompanied by factory representatives, instructors, and evaluators to assist in the training. Reactions of the two units to the Sheridans were quite dissimilar, and in large part affected their approach to training. The 3d Squadron, 4th Cavalry, had elected not to stand down an entire troop for transition; as a result key leaders were often absent from briefings and training. The 11th Armored Cavalry, on the other hand, gave an entire troop seven days of uninterrupted training. In addition, in the 3d Squadron, 4th Cavalry, the Sheridan replaced the M48 tank in cavalry platoons one for one, while in the 11th Armored Cavalry three Sheridans replaced two cavalry platoon armored cavalry assault vehicles. Thus, one unit exchanged a somewhat less capable armored vehicle for its M48 tank, while the other unit exchanged two lightly armed and armored cavalry assault vehicles for three Sheridans—armored vehicles with considerably greater firepower and armor.

On 15 February 1969, in the first combat action involving a Sheridan, one from the 3d Squadron, 4th Cavalry, struck a 25-pound pressure-detonated mine. The explosion ruptured the hull and ignited the combustible-case ammunition of the main gun, causing a deadly second explosion that destroyed the vehicle.

Sheridan crews were uneasy after this catastrophe; they knew that a similar explosion under an M48A3 tank would simply have blown off a few road wheels. The feeling that the Sheridan was extremely dangerous began to grow, and, in the manner of any rumor, spread from unit to unit in Vietnam, and even reached the training base in the United States.

After the mine incident, the effectiveness of the Sheridan was continually suspect in the 4th Cavalry. Then, on 10 March 1969, in a night bivouac at a road injunction east of Tay Ninh City, a Troop A listening post reported enemy movement and the troop went to full alert. Sheridan crews used night observation devices to scan the battlefield. Observing a large group of advancing North Vietnamese, the Sheridans fired canister into the enemy ranks. Confused by the overwhelming volume of fire, the North Vietnamese broke and ran. The next morning more than forty enemy dead, including a battalion commander and a company commander, were found on the battlefield. Reports of this action quickly spread through the squadron, restoring some measures of confidence in the Sheridan.

In contrast the 11th Armored Cavalry's first combat with the Sheridan was successful. In early February 1969, anticipating an enemy offensive, the regiment's 1st Squadron moved to Bien Hoa as a reaction force. Task Force Privette, commanded by Major William C. Privette, the squadron executive officer, included Troops A and B of the 1st Squadron. After an enemy mortar and rocket attack on 23 February, Task Force Privette moved out on an armored sweep and immediately encountered an enemy force. Placing the Sheridans on line, the two cavalry troops moved forward, firing canister into the enemy ranks. In the face of this firepower, the Viet Cong panicked and fled, leaving behind over eighty dead. This fight demonstrated the devasting effect of the 152-mm. canister round. The troops were impressed with the Sheridan's firepower as compared with that of the armored cavalry assault vehicle.

By the end of the test period, both units had concluded that the Sheridan had greater mobility, firepower, range, and night-fighting ability than the vehicle it replaced. On the strength of this conclusion, more Sheridans were sent to Vietnam, and the total number had increased to more than 200 by late 1970. Eventually almost every cavalry unit in Vietnam was equipped with the Sheridan, but the fact remained that the Sheridan's combustible-case ammunition could be detonated by a mine blast or a hit by a rocket propelled grenade. Consequently, the crew of a Sheridan abandoned it quickly after a hit; in contrast, the crew of an M48A3 tank could and did stay and fight after several hits. Another disadvantage was that during the wet season, when vehicles were drenched every day, the Sheridan's electrical fire-control system broke down repeatedly.

"Pile-on"

Changing strategy on both sides increased the use of all armored units, especially armored cavalry. The ability to move men and vehicles rapidly into battle was ideal for small, widely separated, independent engagements. Cavalry could move quickly and bring heavy firepower to bear at critical points. Once the enemy was located and the cavalry unit engaged, reinforcements were immediately sent in to prevent the enemy from escaping, then maximum firepower was brought to bear. Rapid reinforcement of a unit in combat was nicknamed "pile-on." In this period of widespread small actions, some form of pile-on became the usual mode of operation; it was well illustrated by the action of the 3d Squadron, 5th Cavalry, during the battle of Binh An.

In June 1968 this squadron was performing reconnaissance missions under operational control of the 1st Cavalry Division in the I Corps Tactical Zone. During one such mission, Troop C, 3d Squadron, 5th Cavalry, with Troop D, 1st Squadron, 9th Cavalry (Air)—the dismounted ground troop of the air cavalry squadron of the 1st Cavalry Division—had advanced from the northwest to within 150 meters of the village of Binh An, thirteen kilometers north of Quang Tri City on the South China Sea. Suddenly, small arms fire and rocket propelled grenades showered the American forces as several North Vietnamese soldiers withdrew into the village. Both troops began firing to maintain pressure on the enemy, while scout sections from Troop C swung to the north and south of the village to cut off the escape routes. Hundreds of civilians fled from the village as Lieutenant Colonel Hugh J. Bartley ordered Troops A and B to reinforce the attacking units and start the pile-on. Shortly, thereafter, a captured North Vietnamese soldier reported that the 300-man K14 Battalion of the 812th North Vietnamese Regiment was dug in at Binh An. Realizing that he now had an enemy battalion with its back to the sea, Colonel Bartley acted quickly. Troop B was ordered to positions north of Binh An. Troop C moved into the center of a horseshoe-shaped cordon along with Troop D, 1st Squadron, 9th Cavalry. By 1030 the four cavalry troops were in position around Binh An. The South China Sea blocked the enemy's escape east, and a Navy Swift boat, a small

PILE-ON OPERATION IN I CORPS, JUNE 1968. *ACAV's and tanks of Troop B, 3d Squadron, 5th Cavalry, attack Binh An.*

coastal patrol craft, was also summoned to seal the seaward escape routes.

Colonel Bartley's requests for fire support brought tactical aircraft, aerial rocket artillery, and 105-mm. artillery. The cruiser *Boston* and destroyers *O'Brien* and *Edson* took station offshore. When Colonel Bartley gave the order to open fire, the area inside the cordon erupted as hundreds of shells crashed in on the target. A naval observer reported the shelling to be so fierce that North Vietnamese soldiers could be seen diving into the sea to escape.

In order to strengthen the cordon and complete the pile-on, Colonel Bartley requested the airlift of two infantry companies from the 1st Cavalry Division. The two companies arrived early in the afternoon: Company C, 1st Battalion, 5th Cavalry, reinforced Troop B on the north side, while Company C of the 2d Battalion joined Troop A on the south. The supporting fire continued for the rest of the afternoon, and was lifted only long enough for a psychological operations team to fly over Binh An, urging soldiers to surrender. There was no response, and the shelling was resumed.

To prevent the North Vietnamese from escaping by night, the enemy command structure had to be broken up. Colonel Bartley ordered Troops C and D to attack toward the sea. The cavalrymen assaulted the village but were stopped short by an impassable drainage ditch, covered by enemy fire. Troop B, with its attached infantry dispersed between the tracked vehicles, then moved out on line to attack the village from the north. To allow Troop B to use all weapons to its front, Troop A soldiers on the south side of the cordon climbed inside their armored vehicles. Troop B swept forward until its fire began to ricochet off the Troop A vehicles, then turned around and fought its way back to its original blocking positions. Colonel Bartley then called for resumption of supporting fire.

The attack of Troop B apparently ended any intention the enemy had of a mass breakout through the cordon. Thereafter, only small groups or individuals tried to escape by sea; tank searchlights illuminated the beaches, exposing the fugitives. Along the inland sides of the cordon, troops using night vision devices between flares occasionally spotted North Vietnamese groping through the dark. Small arms fire stopped them or drove them back. Artillery rounds continued to explode in the village all night.

Morning brought an increase in the shelling, and when the fire was lifted the entire cordon tightened toward the center of Binh An. A short time later the final attack by Troop B was met by no more than scattered enemy resistance. Stunned North Vietnamese soldiers with hands held high began to stumble from the wreckage toward the American forces. As the search of the village progressed, it became apparent that the K14 Battalion had been eliminated. Over 200 bodies were found and 44 prisoners were taken. Among the dead were the battalion commander, his staff, all the company commanders, and the regimental S–1. Three American soldiers had been killed. The executive officer of the 3d Squadron, 5th Cavalry, Major Michael D. Mahler, writing several years later of the fight at Binh An, stated:

> We had once more stumbled into a situation and been able to turn it to our advantage. But it was more than stumbling and it was not luck that brought success. It was soldiers in hot steel vehicles out in the glaring sand looking and poking until the enemy, North Vietnamese and Viet Cong, never knew when or where an armored column would crop up next.

Rome Plows

Not every armored operation carried with it the excitment and clear victory of a well-executed pile-on. Many, such as maintaining route and convoy security, for example, involved only occasional meetings with the enemy, but were hard work and were often boring. One operation that almost every armored unit participated in at one time or another was the protection of engineer units clearing large areas of jungle with heavy Rome plows. Usually such operations extended over a considerable time and involved as many as fifty or more plows. First tried in the III

ROME PLOWS WITH SECURITY GUARD OF M113'S

Corps Tactical Zone, Rome plows soon became commonplace in the two corps zones to the north, although they were not useful in the delta. Some jungle clearing was done during operations such as CEDAR FALLS and JUNCTION CITY; however, major land-clearing efforts did not begin until the arrival of the 169th Engineer Battalion in May 1967.

The task of protecting the clearing operations created a need for techniques for which there were no precedents. One method was developed during the summer of 1967 by the 4th Battalion, 23d Infantry (Mechanized), which cleared the Iron Triangle in III Corps, an area of thick undergrowth and trees of small to medium size. Daily operations were carried out by three land-clearing teams, each composed of eight Rome plows and two conventional bulldozers, a security force, and a combined arms force for search and reaction missions. The security force usually consisted of a mechanized infantry company (minus one rifle platoon) and a tank platoon. Because enemy base camps, tunnels, and other installations were frequently uncovered, each security force contained a search group of one infantry platoon and an engineer squad. When an enemy base camp was discovered, the clearing team went on working while the search group gathered information and enemy materiel before destroying the camp. The combined arms force worked with the same plow team throughout the operation in order to insure close teamwork.

Before the plows began clearing, preparatory machine fire, mortar, and tank canister fire was directed into the jungle. Once the first swath, outlining the area to be cleared, was completed the security force deployed in single file outside each plow team, with a tank section at the front and rear of the column. To discourage ambushes, harassing fire was constantly delivered on the uncleared jungle surrounding the working area. If enemy soldiers were discovered, the security force immediately deployed and assaulted while one platoon escorted the plows to safety. To prepare night defenses, plow teams cleared firing lanes and dug positions for the tracked vehicles.

In Rome plow operations along Route 20 from Blackhorse Base Camp to the boundary between III Corps and II Corps, the 3d Squadron, 11th Armored Cavalry, used a different procedure. Instead of keeping the bulk of its cavalry with the plows, the squadron first cleared the immediate area to be cut, then left a security force of two to four armored cavalry assault vehicles with the plows. The remainder of the squadron conducted search and clear operations around the area being worked to keep the enemy from entering. This method became more common in later years. As emphasis on clearing along major roads and in enemy base areas increased, Rome plow security operations became routine for armored units. In the absence of established tactics, units used methods whose variety was limited only by the ingenuity of the commanders.

Tank Versus Tank

Combat between tanks, for which U.S. tank crews traditionally train, materialized only once for American armored units in Vietnam. Since the North Vietnamese had armored forces in October 1959, when the 202d Armored Regiment was created, their reasons for not making more use of them earlier can only be conjectured.[8] Cadre from the regiment had joined the Central Office for South Vietnam staff as armor advisers as early as 1962. The North Vietnamese Army established an armor command in the summer of 1965, and by March 1973 there were more than twelve armor units, ranging in size from battalion to regiment. The People's Republic of China and the Soviet Union supplied all the armored vehicles, which included armored personnel carriers, light tanks, and medium tanks. The associated armored support vehicles, including self-propelled antiaircraft weapons, were supplemented by captured U.S. M41 tanks. Although poorly managed in early encounters,

[8] The cadre of this regiment studied tank warfare in Communist China and the Soviet Union in the mid-1950's. The numerical designation of the regiment was derived from the 202 cadre members who had been trained abroad.

enemy armored forces learned their lessons well, and in early 1975 actually spearheaded the final assaults on South Vietnam.

Contrary to American teaching on the subject, the North Vietnamese Army did not advocate the use of tanks in mass. Its doctrine stated that armor would be employed during an attack, when feasible, to reduce infantry casualties; however, only the minimum number of tanks required to accomplish the mission would be used. Battle drill dictated that lead tanks were to advance, firing, and to be supported by fire from other tanks and from artillery. Close coordination between tanks and supporting infantry was stressed as a key to success in the attack. Because the North Vietnamese lacked air power, they placed strong emphasis on camouflage training in armor units; in the spring offensive of 1972, tank regiments moved great distances without being detected.

Until the end of 1973, North Vietnamese armor appeared on or near the battlefields of South Vietnam on only four recorded occasions. The first instance was at Lang Vei Special Forces Camp near Khe Sanh in the I Corps area on 6–7 February 1968. Here, a North Vietnamese combined arms attack with PT76 tanks succeeded in breaking through the camp's defensive positions. In 1969 at Ben Het the North Vietnamese again used armor. Against the South Vietnamese Army force that attacked into Laos in 1971, the North Vietnamese committed an entire armor regiment and staged well-coordinated tank-infantry attacks. In their spring offensive of 1972 in South Vietnam, the North Vietnamese used the largest tank forces of the war. Entire tank companies stormed objectives, with infantry troops following close behind.

It was at Ben Het in March 1969 that American and North Vietnamese armor clashed for the first and only time. The Ben Het Special Forces Camp in the central highlands of the II Corps Tactical Zone overlooked the Ho Chi Minh Trail where the borders of Laos, Cambodia, and Vietnam come together. In an effort to mask nearby North Vietnamese troop movements, the enemy had subjected the camp to intense indirect fire attacks during February.

To counter an apparent enemy buildup, elements of the 1st Battalion, 69th Armor, were sent to the area.[4] Captain John P. Stovall's Company B, the forward unit of the battalion, occupied strongpoints and bridge security positions along the ten-kilometer

[4] In November 1968 a pilot from the 7th Squadron, 17th Cavalry (Air), reported four unidentified tanks in this area but the report was never confirmed. The 4th Infantry Division had reports from various sources that enemy tanks were present in the area.

road link between Ben Het and Dak To. One platoon of tanks was stationed in the camp. Free world forces at Ben Het included three Vietnamese infantry companies and their Special Forces advisers, an American 175-mm. artillery battery, and two M42's, tracked vehicles mounting 40-mm. twin guns on an M41 tank chassis. The M42's and the 175-mm. battery were in the main camp, while most of the newly arrived tank platoon took up dug-in positions on a hill facing west toward Cambodia. One tank, located in the main camp, occupied a firing position guarding the left flank overlooking the resupply route. Through February the platoon endured heavy enemy shelling by taking cover in its armored vehicles and moving from bunker to bunker during quiet periods. The crews fired their 90-mm. guns at suspected North Vietnamese gun sites and bunkers on the rugged slopes. When the tank platoon leader was wounded and evacuated, Captain Stovall moved the company command post to Ben Het.

Enemy shelling decreased in March, allowing defensive positions to be strengthened and improved, and making the entire camp ready for instant action. For three full days enemy fire abated, but at 2100 on 3 March 1969 the camp once again began to receive mortar and artillery fire in crashing volleys. Both Sergeant First Class Hugh H. Havermale and Staff Sergeant Jerry W. Jones heard the sound of tracks and heavy engines through the noise of the artillery. With no free world tanks to the west, the probability of an enemy tank attack sent everyone into action. High explosive antitank (HEAT) ammunition was loaded into tank guns and from battle stations all eyes strained into the darkness.

In his tank, Sergeant Havermale scanned the area with an infrared searchlight, but could not identify targets in the fog. Sergeant Jones, from his tank, could see the area from which the tank sounds were coming but had no searchlight. Tension grew. Suddenly an antitank mine exploded 1,100 meters to the southwest, giving away the location of the enemy; the battle for Ben Het now began in earnest.

Although immobilized, the enemy PT76 tank that had hit the mine was still able to fight. Even before the echo of the explosion had died, the PT76 had fired a round that fell short of the defenders' position. The remainder of the enemy force opened fire, and seven other gun flashes could be seen. The U.S. forces returned the fire with HEAT ammunition from the tanks and fire from all other weapons as well. Specialist 4 Frank Hembree was the first American tank gunner to fire, and he remembers: "I only had his muzzle flashes to sight on, but I couldn't wait for a better target

RUSSIAN-MADE PT76 TANK DESTROYED AT BEN HET

because his shells were landing real close to us." The muzzle flashes proved to be enough for Specialist Hembree; his second round turned the enemy tank into a fireball.

Capital Stovall called for illumination from the camp's mortar section and in the light of flares spotted another PT76. Unfortunately, the flares also gave the North Vietnamese tanks a clear view of the camp's defenses, and as Captain Stovall was climbing aboard Sergeant Havermale's tank, an enemy high explosive round hit the loader's hatch. The concussion blew Stovall and Havermale from the tank, and killed the driver and loader. Damage to the tank was slight.

Sergeant Jones took charge, dismounted, and ran to another tank which was not able to fire on the enemy main avenue of approach. Still under hostile fire, he directed the tank to a new firing position where the crew quickly sighted a PT76 beside the now burning hulk of the first enemy tank. The gunner, Specialist 4 Eddie Davis, took aim on one of the flashes and fired. "I wasn't sure of the target," Specialist Davis said, "but I was glad to see it explode a seond later." Every weapon that could be brought to bear on the enemy was firing. Having exhausted their basic load of high explosive antitank ammunition, the tank crews were now firing high explosives with concrete-piercing fuzes. Gradually, the enemy fire slackened, and it became clear that an infantry assault was not imminent. In the lull, the crews scrambled to replenish their basic load from the ammunition stored in a ditch behind the tanks. Tank rounds were fired at suspected enemy locations but there was no return fire. The remainder of the night was quiet; the tension of battle subsided, and the wounded were evacuated.

The battle for Ben Het had not gone unnoticed by the remainder of the 1st Battalion, 69th Armor. Company A and the battalion command post moved to Polei Kleng to reinforce ground elements and be in a position to counterattack population centers. The 2d Platoon of Company B assembled and moved by night to Ben Het, where a search of the battlefield the next day revealed two PT76 hulls and an enemy troop carrier that had not been noticed during the battle but now lay burned out and abandoned on the edge of the battlefield. The enemy vehicles were part of the 16th Company, 4th Battalion, 202d Armored Regiment of the North Vietnamese Army.

Intelligence later revealed that the main object of the attack on Ben Het was to destroy the U.S. 175-mm. guns. Whatever the enemy's intention, the camp was held by American tanks against North Vietnamese tanks. Not until March 1971, when South Vietnamese M41 tanks battled North Vietnamese tanks in Laos, would tanks clash again.

Invading the Enemy's Sanctuaries

The battle that ended in the defeat of the enemy at Ben Het was only one of an increasing number of attacks in which the enemy did not achieve military victory. Free world forces were no longer content to sit back and wait for North Vietnamese troops to make the first move. Base areas, once safe havens for the enemy, were penetrated by large armored formations intent on disrupting the enemy logistical system. General Abram's strategy of destroying the enemy logistics nose was now in full swing.

One of these operations was conducted during March and April 1969 by elements of the 1st Brigade, 5th Infantry Division in western Quang Tri Province of the I Corps Tactical Zone. Operating in country long thought to be impenetrable to armored vehicles, this combined arms team, designated Task Force REMAGEN, demonstrated again the advantage of mechanized forces. Of special significance during this operation was the lack of a ground line of communications to the more than 1,500 men of Task Force REMAGEN. Helicopters supplied the task force with over 59,000 gallons of diesel fuel and gasoline and more than 10,000 rounds of 105-mm. artillery ammunition. Although the task force encountered normal maintenance problems as it moved through the rough terrain, tank power packs weighing over four tons and other major components were delivered by helicopter. For forty-three days, the task force operated in rugged terrain along the Laotian border on an aerial supply line, demonstrating that even remote base areas were vulnerable to attack by armored units.

The success of REMAGEN was not an isolated case, for the feat was duplicated in the III Corps Tactical Zone. Operation MONTANA RAIDER, conducted from 12 April to 14 May 1969 in the area east and north of Tay Ninh City, was aimed at a rear service support and transportation zone for enemy troops and equipment entering South Vietnam from Cambodia. Although the exact location and identity of enemy units in this region were not known, two North Vietnamese divisions were thought to be present. The terrain was not rugged, but dense jungle hampered movement. The MONTANA RAIDER force consisted of one infantry-heavy and two armor-heavy task forces under command of the 11th Armored Cavalry. The regiment's air cavalry troop and the 1st Squadron, 9th Cavalry, the air cavalry squadron of the 1st Cavalry Division, flew in support of the operation. An artillery battalion headquarters under direct control of the regiment coordinated all artillery fire. A cover and deception plan was devised to persuade the enemy that U.S. forces were moving north and west of Tay Ninh City. Air cavalrymen, flying over enemy base camps, deliberately lost map overlays clearly marking the area northwest of Tay Ninh City as an objective for the operation, and intentional security breaches in radio transmissions were employed to the same end.

At 0800 on 12 April operational control of the 11th Armored Cavalry passed from the 1st Infantry Division to the 1st Cavalry Division, marking the beginning of MONTANA RAIDER. In accordance with the deception plan, the armor-heavy task forces left Bien Hoa and moved past the actual area of operations. As the 11th Armored Cavalry's 2d Squadron task force neared Dau Tieng, it swung northwest to join Company A of the 1st Battalion, 8th Cavalry (Airmobile), while the 11th Cavalry's 1st Squadron task force moved into another base and linked up with Company C of the 1st Battalion, 8th Cavalry. The movement through and beyond the actual area of operations was designed to suggest further to the enemy that the operation would be conducted northwest of Tay

Ninh City. By 1700 on 12 April all forces had completed the 98-kilometer move and were ready for action.

On 13 April Colonel James H. Leach, commander of the 11th Armored Cavalry took operational control of an airmobile infantry unit, the 1st Battalion, 8th Cavalry, and began reconnaissance in force operations east to Tay Ninh City. The 11th Cavalry's 1st Squadron task force entered the area from the southwest, its 2d Squadron task force from the northwest, and the 8th Cavalry task force from the northeast. In order to give the 8th Cavalry task force additional firepower and some armored protection, Troop G and one platoon of Company H, the tank company of the 2d Squadron, 11th Armored Cavalry, were attached. The first unit to clash with the enemy was the regimental air cavalry troop, which was assessing bomb damage from a B–52 strike. After the aerorifles and infantry reinforcements were sent in, Troop A of the 1st Squadron, 11th Armored Cavalry, arrived, and in an all-day battle in the heavy jungle finally drove the enemy out. The following days saw scattered fighting as the task forces converged. Artillery and air strikes were used liberally to destroy enemy base camps. The longest battle of Phase I occurred on 18 April when Troops A and B of the 1st Squadron, 11th Armored Cavalry, met a large enemy force. Heavy artillery and air strikes were used against the enemy, but the assault was delayed when machine gun tracer ammunition created a fire storm in the bamboo thickets. The enemy lost seventy-six men in this battle.

At the end of Phase I, and after two days devoted to maintenance, Phase II opened with a 149-kilometer road march for the entire regiment to Quan Loi in Binh Long Province, 100 kilometers north of Saigon. Phases II and III saw the combined arms task forces of the 11th Armored Cavalry ranging throughout eastern War Zone C, engaging the enemy in short, bitter fights, almost always in heavy jungle. The stress again was on mobility, firepower, and the combined arms team.

MONTANA RAIDER demonstrated the versatility of a large, mounted unit, aggressively led and employing conventional armored award doctrine in isolated jungle. All three phases of MONTANA RAIDER again showed the value of combined arms—armored cavalry, infantry, artillery, and air cavalry. Surprising mobility was achieved by tracked vehicles, which covered more than 1,600 kilometers during the operation; of that distance, 1,300 kilometers were in dense jungle. More important was the fact that this operation, REMAGEN in the north, and others throughout Vietnam put free world forces in possession of the enemy base areas during 1969. With nowhere else to go, the Viet Cong and North Vietnamese pulled back to their bases in Cambodia and Laos.

In the southern part of I Corps Tactical Zone, in the spring of 1969, a similar base area operation was conducted in the A Shau valley along the Laos border. Spearheaded by the 101st Airborne Division, it included eighty tracked vehicles of the 3d Squadron, 5th Cavalry, and the 7th Vietnamese Armored Cavalry Regiment. These were the first armored forces to operate in the A Shau valley.

Securing the Borders

After the success of operations against enemy sanctuaries in South Vietnam, the next step was sealing the borders, or at least making them reasonably secure. With the growing demands of pacification and the prospect of troop withdrawals, which would limit the resources available, the task naturally fell to mobile units, both ground and air, that could move rapidly and control large areas. Armored and airmobile units became the mainstay of border operations, particularly those in the critical III Crops area north of Saigon.

Based at Phuoc Vinh, the 1st Cavalry Division, with three brigades of airmobile infantry and operational control over the 11th Armored Cavalry, was extended among more than a hundred kilometers of border from east of Bu Dop to northwest of Tay Ninh City, opposite enemy base area 354. The 25th Infantry Division controlled the western and southern approaches to Saigon, and the 1st Infantry Division commanded the entrances to the Saigon River corridor and the old, now quiet war zones in southern Binh Long and Phuoc Long provinces.

Border operations of the armored cavalry, the air cavalry, and the airmobile infantry of the 1st Cavalry Division illustrate the tactics of both sides in the conflict. As the enemy tried to cross the border in strength, supported from bases beyond South Vietnamese boundaries, the defenders attempted to prevent the crossing with firepower and maneuver. By early 1970 the cavalry and airmobile infantry forces had developed some sophisticated techniques employing Rome plow cutting, sensors, and automatic ambush devices to deny the use of trails to the enemy. These techniques were first applied systematically in the northwestern part of the III Corps area from Bu Dop to Loc Ninh along QL-14A, and almost immediately produced good results. (*See Map 13, inset.*) The 2d Squadron, 11th Armored Cavalry, controlled this region from Fire Support Base Ruth outside Bu Dop and patrolled east and west along the Cambodian border.

Early on the morning of 20 January 1970, as Lieutenant Colonel Grail L. Brookshire, commander of the 2d Squadron, and Colonel Donn A. Starry, the regimental commander, conferred at Fire Support Base Ruth, a deluge of mortar and rocket fire descended on the base. Both commanders took off in helicopters, and while Colonel Brookshire gathered his squadron Colonel Starry requested tactical air, air cavalry, and artillery fire support. Two battalions of the 65th North Vietnamese Regiment and part of an antiaircraft regiment had occupied a dry lake bed about three kilometers west of the base; it was from near this crescent-shaped opening in the otherwise dense jungle that the indirect fire attack on the base began. *(Map 13)* Later it was learned that the enemy had hoped to lure the American forces into an airmobile assault into the clearing, where carefully sited antiaircraft guns would have devastated such a force.

As gunship, tactical air, and artillery fire was brought in, a scout helicopter was shot down, leaving the wounded pilot stranded in a bomb crater. The 2d Squadron began to move to the location, with the tanks of Company H approaching from the north and the cavalry of Troops F and G from the south. A Cobra pilot, Captain Carl B. Marshall, located the flaming wreckage and spotted the wounded pilot, First Lieutenant William Parris, waving from the bomb crater. Captain Marshall flew in low and landed nearby in a hail of machine gun and mortar fire. Lieutenant Parris raced to the aircraft and dove into the front seat, where he lay across the gunner's lap, legs danging from the open canopy, as Captain Marshall pulled up, barely clearing the wood line.

Artillery, fighter bombers, and gunships descended on the enemy, while Troop F hastened up Highway 14A to link up with Troop G, which was already heading north. After crossing a meadow near Bau Ba Linh and a seemingly unfordable stream, Troop G bore into the jungle. Ninety minutes and two kilometers of single and double canopy jungle later, Troop G arrived, joined Troop F, and, on line, the cavalry assaulted. From the helicopter Colonel Starry had meanwhile called for an airdrop of tear gas clusters on enemy bunkers toward the north side of the crescent. Colonel Brookside halted the artillery fire long enough for the drop, which brought the enemy troops out of the bunkers and sent them running north for the border. Colonel Brookshire then ordered artillery and gunship fire while Troops G and F attacked through the enemy positions. Company H, in position north of the crescent, caught the fleeing enemy with canister and machine gun fire.

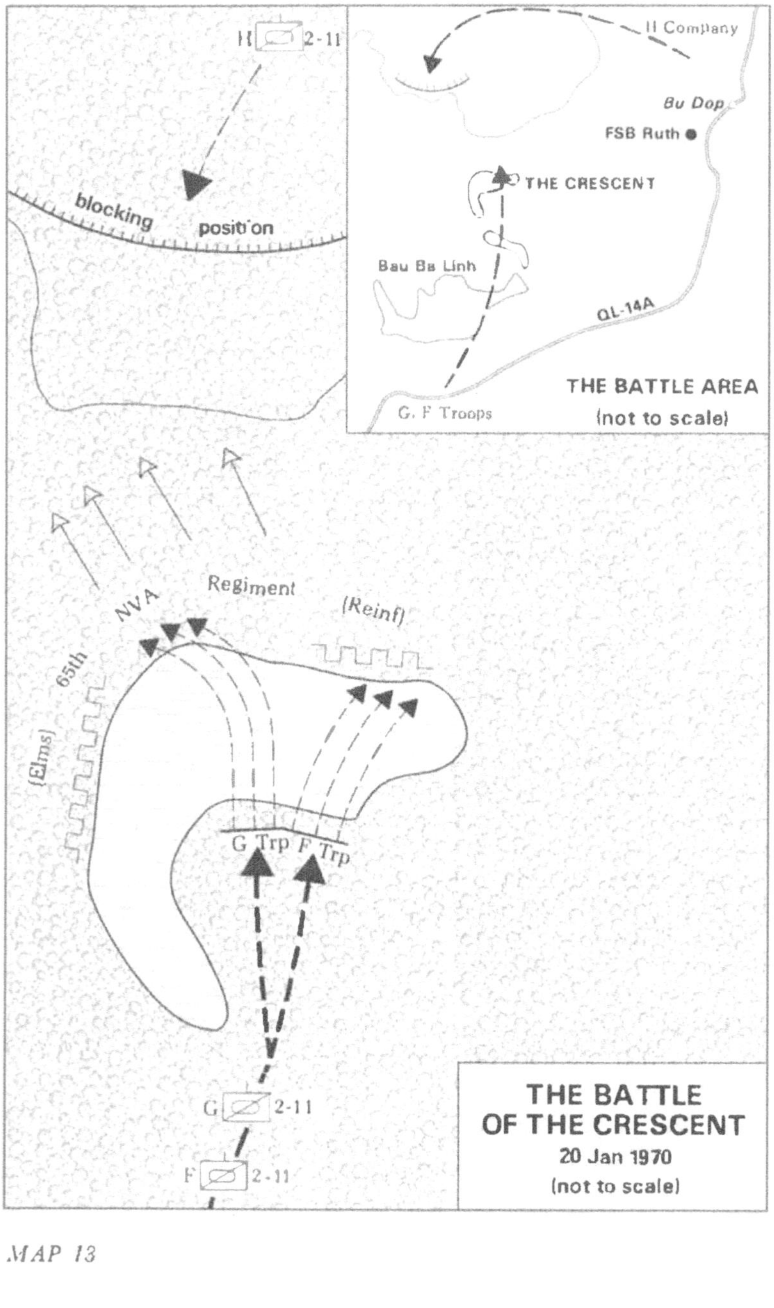
H 2-11
blocking position
H Company
Bu Dop
FSB Ruth
THE CRESCENT
Bau Ba Linh
QL-14A
G, F Troops
THE BATTLE AREA
(not to scale)
Regiment
NVA
(Reinf)
65th
(Elms)
G Trp
F Trp
G 2-11
F 2-11
THE BATTLE
OF THE CRESCENT
20 Jan 1970
(not to scale)

MAP 13

The ground forces continued to fight and maneuver until nightfall. Darkness prevented a detailed search, and the next morning the 2d Squadron pulled out at dawn. The fight had lasted almost fourteen hours, with over 600 rounds of artillery fired, thirty tactical strikes employed, and fifty Cobra rocket loads delivered. The 65th North Vietnamese Regiment did not appear again in battle for nearly four months. Colonel Brookshire's search of the crescent was broken off abruptly as he moved to reinforce the regiment's 1st Squadron, fighting near An Loc.

Early on 21 January 1970, thirty-five kilometers to the west, the 1st Squadron, which was operating in a Loc Ninh rubber plantation, intercepted two North Vietnamese battalions moving south into Loc Ninh District. The 11th Armored Cavalry piled on with Colonel Brookshire's 2d Squadron racing in from the crescent battleground to the northeast and Lieutenant Colonel George C. Hoffmaster's 3d Squadron attacking north along Highway 13 out of An Loc. Over half of the regiment converged on the fight in less than three hours and broke the back of the enemy attack. The two North Vietnamese battalions ran for cover in Cambodia, with elements of the cavalry pursuing them to the border. The pursuit was aided by a map found on the North Vietnamese commander that showed the escape plan. Those fugitives who reached the escape routes were met by tactical air and artillery fire. After these two fights and a few more with the same outcome, the enemy showed reluctance to risk a fight with the cavalry, whose mobility and firepower had been overwhelming.

Operations along Highway 14A were so successful in drying up enemy logistical operations along the jungle trails that it was decided to repeat the scheme in War Zone C. The plan was to stop the supply operations of the enemy's 50th Rear Service Group operating out of the Cambodian Fishhook area into the Saigon River corridor through War Zone C. The American forces were to conduct extensive land clearing operations along Route 246, generally east and west across War Zone C, thus blocking the north and south trails from Cambodia to the Saigon River corridor.

By mid-February Rome plows had cleared a swath of jungle 400 to 500 meters wide along Highway 246 from An Loc in Binh Long Province to Katum in Tay Ninh Province, just south of the border. Along this cut the 11th Armored Cavalry began operations. The 3d Squadron anchored the east flank near An Loc, the 2d Squadron held the middle, and the 1st Squadron covered the west in northwestern Tay Ninh Province. Airmobile infantry battalions of the 1st Cavalry Division, operating south in War Zone C, frequently under control of the 11th Armored Cavalry, completed the interdiction force.

Colonel Starry was convinced that to cut enemy supply lines successfully ground had to be held, and that control of the ground followed from constant use of the ground. The operational pattern of the regiment, therefore, was one of extensive patrolling, day and night, and the setting up of an intricate network of manned and unmanned ambushes all along the trail system. The cavalry soon came to know the enemy's trails well, and by clever use of automatic devices reduced enemy logistical operations to a trickle. The ambush net cost the enemy ten to thirty casualties each night. Every site was checked, and electrical devices were moved and reset each day. It was like running a long trapline.

Monitoring enemy radio traffic, the 11th Armored Cavalry learned that enemy units to the south were desperate for food and ammunition. Enemy relief parties were killed in ambushes or by cavalry units that took advantage of information gleaned from careless enemy radio operators. Enemy messengers sent along the trails were killed or captured; their messages and plans provided information for setting up more traps.

The enemy, reluctant to confront the cavalry directly, attacked only by fire in War Zone C, and tried to outflank the net of ambushes. The 209th North Vietnamese Army Regiment lost over 200 men when it ran headlong into Captain John S. Caldwell's Troop L, 3d Squadron, in the Loc Ninh rubber plantation in March. Later, in April, the 95C North Vietnamese Regiment, trying to move west around the ambush system, encountered Lieutenant Colonel James B. Reed's 1st Squadron near Katum. With the cavalry and tanks of the 1st Squadron heavily engaged, Colonel Starry alerted Colonel Brookshire to move two troops from the 2d Squadron west to join in the fight. The enemy now had two battalions locked in combat with the 1st Squadron, while a third battalion was escaping to the north. Realizing he faced the cavalry regiment, the enemy commander panicked and began broadcasting instructions to his battalions in the clear. As the enemy troops tried to disengage, intercepts of the instructions they were receiving were passed to Colonel Reed. Armed with this information, the 1st Squadron blocked the enemy. In the ensuing melee the cavalry squadrons virtually destroyed the two battalions opposing Colonel Reed. Some of the third battalion to the north escaped despite air strikes and artillery fire placed along the escape routes.

It was more than six months before the North Vietnamese 95C Regiment fought again. The extensive system of Rome plow cuts and the presence of cavalry and airmobile forces in late 1969 enabled free world forces to choke off enemy supply lines and neutralize bases in War Zone C. More so than at any other period in the war, except when the attacks were made into Cambodia, enemy access to South Vietnam was cut off.

Pacification Efforts

While operations against the sanctuaries and along the border were in progress, important steps were taken to win over the people in areas long under enemy control. In the spring of 1969 major enemy action had diminished to a level that offered an opportunity for large-scale efforts in this direction. On 13 April 1969 the 173d Airborne Brigade began Operation WASHINGTON GREEN, which would occupy the brigade for the next nineteen months. The mission called for placing one mechanized and three infantry battalions in four densely populated districts of Binh Dinh Province in the II Corps Tactical Zone. Primary emphasis was to be given village and hamlet protection in order to enable territorial forces, in conjunction with other government agencies, to conduct searches behind a protective shield of South Vietnamese Army and American forces. U.S. troops were usually present at first to supplement weak local security forces until recruitment and training would permit replacement of American by Vietnamese units. The 1st Battalion, 50th Infantry (Mechanized), was assigned to Phu My District for most of this operation.

Even large units such as the 1st Brigade, 5th Infantry Division (Mechanized), became totally involved. In January 1970 the brigade launched Operation GREEN RIVER, designed to further the pacification program in the northern I Corps area by conducting combined operations with the South Vietnamese 1st Infantry Division. Reconnaissance in force, search and clear operations, and measures against enemy rockets were undertaken throughout Quang Tri Province.

By 1970 many American units were committed to securing large areas in the interest of the South Vietnamese government's pacification program, and some armor units had that as a primary mission. Extensive surveillance operations were conducted along the Demilitarized Zone to prevent enemy infiltration, and security screens were established around populated and militarily significant areas.

Forces in Operation GREEN RIVER killed over 400 of the enemy in six months. As GREEN RIVER ended, units of the brigade began Operation WOLFE MOUNTAIN, which lasted into January 1971.

A Tank of the 2d Battalion, 34th Armor, in Position to Provide Static Road Security

Combined operations were conducted by the brigade, the South Vietnamese 1st Infantry Division, and territorial forces throughout northern Quang Tri Province. The operations included the use of armored forces as security along the Demilitarized Zone, on lines of communication, and around populated areas.

Pacification efforts and measures to aid the Vietnamese armed forces to assume the full burden of the war intensified in mid-1969 with U.S. troop withdrawals. The South Vietnamese Army undertook a program to expand the armored cavalry from ten to seventeen regiments. Even more noteworthy was the activation of two armor brigade headquarters, which allowed South Vietnamese armored units to operate in larger formations. Both the I and IV Armor Brigades deployed to their respective corps tactical zones during 1969, and were soon followed by two more brigades. Each armor brigade was a highly mobile, independent, tactical headquarters that could control ten to twelve squadrons.

On 22 and 23 May 1969, a joint Vietnamese-American armor conference convened: attending were the South Vietnamese Chief of Armor, with his staff and his American advisers, and all South

Vietnamese armor regimental commanders and their American advisers. Their purpose was to review South Vietnamese armor and set goals for its future development. The key question was whether current missions were making full use of armored units. The general response was no. Fragmenting of armored units, static missions, the use of tanks as pillboxes, and assigning armored forces permanent areas of operation were the most common mistakes singled out by South Vietnamese armor leaders. The South Vietnamese Armor Command made an honest effort to evaluate itself, and took positive action to improve its performance.

In November 1969 the Vietnamese Joint General Staff published a directive on employment of Vietnamese armored units. The directive first noted improper uses that had been described by the armor commanders, and then added that many units had failed to provide logistical support for armored units assigned to them. It directed certain corrective actions.

1. Avoid the use of armored forces in static security missions.

2. Do not divide armored units below troop level.

3. Give missions of reconnaissance and search and destroy in large operational areas.

4. Use armor brigade and regimental headquarters to direct and control combined arms operations.

5. Use armored units in night operations with the support of organic searchlights, mortars, flares, artillery, and aircraft.

6. Develop U.S. and Vietnamese combined operations.

This analysis of the South Vietnamese Army's use of armor and the subsequent directive from the Joint General Staff put backbone into South Vietnamese armor doctrine. Although it ruffled some feelings in the Vietnamese command, improvements in field use were noticed immediately. In February 1970 the 1st Armored Brigade conducted mobile independent operations along the sea in the northern part of the I Corps Tactical Zone. Controlling up to two regiments of cavalry, Rangers, and territorial forces, for two months the brigade roamed over the area and succeeded in destroying three enemy battalions. As part of the operation, 5,000 acres of land were cleared. The enemy was effectively defeated and moved away: Regional and Popular Forces units moved in and established permanent settlements. Almost 900 Viet Cong and North Vietnamese were killed or captured, while the brigade lost sixty-eight men. For success in its first large-scale operation, the Vietnamese 1st Armored Brigade was awarded a U.S. Presidential Unit Citation.

Vietnamese Forces Take Over the War

In the United States, as the newly elected president, Richard M. Nixon, prepared to take office in Washington in late 1968, the single most vexing problem confronting the administration was Vietnam. Unable to resolve the issue satisfactorily, Lyndon B. Johnson had chosen not to seek another term. In response to instructions from Washington, U.S. Ambassador Ellsworth Bunker and General Abrams had privately discussed with President Thieu, the possibility of withdrawing some American forces. By January 1969 these conversations had expanded into specific proposals for sending home first one, then two American divisions. Then, in April, the new administration issued National Security Study Memorandum 36, directing preparation of plans for turning the war over to the Vietnamese.

In Vietnam plans were drawn up in strictest secrecy, under the careful eye of General Abrams himself by a very small task force headed by Colonel Starry. At the outset the idea that withdrawal of a single American soldier would cause the collapse of the whole war effort was, to use the words of General Abrams "simply unthinkable." General Abrams, however, was firmly convinced that the Vietnamese Army could do more. He drew considerable confidence from the growing success of the pacification effort, and, always a practical man, he realized, that like it or not, the new administration was committed to withdrawing some or all American forces. His instructions to Colonel Starry were quite clear: ". . . do it right, do it in an orderly way . . . save the armor units out until last, they can buy us more time." Thus armor units, specifically excluded from the buildup until late 1966, would anchor the withdrawal of American combat units from Vietnam.

On 9 June 1969 President Nixon met President Thieu at Midway Island and they agreed to the first withdrawal of U.S. troops from Vietnam—25,000 men. On 13 July Company C, 3d Marine Tank Battalion, became the first U.S. armor unit to leave South Vietnam as one battalion landing team of U.S. marines boarded its amphibious ships. Troop withdrawals continued at an ever-accelerating pace, even while large-scale operations, such as the incursion into Cambodia of 1970 and the enemy offensive of early 1972 were in progress. From the beginning, force planners held out armored units—tanks, air cavalry, ground cavalry, and mechanized infantry. As divisions or brigades left the country, their armored units remained behind. The mobility and firepower of armored units made them the logical choice for operations over extended areas, and rearguard, delay, and economy of force roles were traditional armor specialties, particularly for cavalry. Thus, when the 9th Infantry Division departed in 1969, the 2d Battalion, 47th Infantry (Mechanized), and the 3d Squadron, 5th Cavalry, remained behind. Almost every air cavalry unit remained in Vietnam until early 1972. These armored units provided a maximum of firepower and mobility with a minimum of U.S. troops. By the end of 1970, with the withdrawal of American units in high gear, fourteen armored battalions or squadrons remained in Vietnam. In December 1971 armored units represented 54 percent of the U.S. maneuver battalions still in Vietnam.

The U.S. armored units that remained supported and trained Vietnamese forces while combat operations were carried out. One such unit, the U.S. 7th Squadron, 1st Cavalry (Air), supported the Vietnamese 7th, 9th, and 21st Infantry Divisions in the delta and along the Mekong River corridor to Cambodia. On occasion, air cavalry units used South Vietnamese troops as aerorifle platoons. In addition, the squadron trained Vietnamese pilots in a program calling for three months or 180 hours of flight time for each pilot. During this successful program, it was found invaluable to have an individual who spoke Vietnamese aboard each American helicopter while the aircraft were supporting South Vietnamese operations. Troop D, the ground troop of the squadron, provided instruction in small unit tactics for Vietnamese Regional and Popular Forces.

CHAPTER VII

Across the Border: Sanctuaries in Cambodia and Laos

As early as 1965 the North Vietnamese used areas of Cambodia and Laos near the borders of South Vietnam as sanctuaries in which to stock supplies and conduct training without interference. It was in these countries that the North Vietnamese built the famous Ho Chi Minh Trail as their principal supply route to the south. As time wore on and the tempo of the war increased, the word trail became a misnomer, for a primitive network of jungle paths had grown into a vast system of improved roads and trails, many of which could be used the year-round. The image of a North Vietnamese soldier-porter trudging south from Hanoi for six months with two mortar shells destined for South Vietnam could no longer be conjured up. By late 1968 the North Vietnamese were moving most of their supplies by truck, pipeline, and river barge.

This relatively sophisticated transportation system terminated at depots within and adjacent to South Vietnam. Combat units in South Vietnam received supplies from these depots by a simpler but highly organized system of distribution that made use of small boats, pack animals, and porters. In the late 1960's, as the free world forces extended their operations into the enemy base areas in South Vietnam, the enemy regular forces expanded the bases and depots across the borders in Cambodia and Laos. (*See Map 14, inset*). Since for political reasons these base areas were inviolate, they provided sanctuaries to which the North Vietnamese and Viet Cong units could retire periodically from combat in South Vietnam, train and refit, and return to combat. Free world forces called these sanctuaries base areas, since they provided not only supply and maintenance facilities but also training and maneuver areas, classrooms, headquarters, and even housing for families of soldiers.

The Cambodian government, under pressure from North Vietnam and China, had for several years conceded these areas to the enemies of South Vietnam. In March 1970, however, Marshal Lon Nol of Cambodia seized control of the government and began a campaign to restrict the Viet Cong and North Vietnamese in their use of his country. Lon Nol's efforts hinged on a proposal that would allow them to continue to use some base areas, but under Cambodian control. Since it would have severely hampered movement of Cambodian-based enemy troops and supplies to and from South Vietnam, the proposal was rejected by the Viet Cong and North Vietnamese, who moved to occupy major portions of eastern Cambodia.

The coalition government of Laos had an arrangement with North Vietnamese sympathizers that did not permit it to object to Viet Cong and North Vietnamese operations in Laos. Base areas and their supporting transportation networks in Laos therefore continued to provide critical support for North Vietnamese forces operating in South Vietnam, and were a thorn in the side to the free world forces. At one time or another most free world soldiers had seen Viet Cong and North Vietnamese troops moving with impunity on the other side of the ill-defined border.

Early Operations Into Cambodia

As the Cambodian situation became worse, the Cambodian government sought military assistance from the United States and South Vietnam. In response the South Vietnamese Army III Corps headquarters launched TOAN THANG 41, a three-day operation into the so-called Angel's Wing, an area in Cambodia long used by the enemy for resting and refitting units. Three South Vietnamese Army task forces, each containing armored cavalry and Ranger units, began the operation at 0800 on 14 April 1970. They were supported in South Vietnam by the U.S. 25th Infantry Division. At midday, eight kilometers inside Cambodia, a sharp fight broke out, and fierce hand-to-hand combat continued until late afternoon, when the enemy broke away and fled. The next day the capture of several base camps revealed the full extent of enemy logistical operations in Cambodia. Plans called for the Vietnamese Air Force to evacuate captured supplies, but because of the inexperience of the Vietnamese in large-scale logistical airlift operations most material had to be evacuated by truck or tracked vehicle. What could not be removed was destroyed. In all, 378 of the enemy were killed and 37 captured. Eight South Vietnamese soldiers were killed. Success brought confidence to the South Vietnamese government and the army.

Thus encouraged, the South Vietnamese Army decided to expand operations into the Crow's Nest area on 20 April 1970. The expedition was planned by the South Vietnamese, and involved three armored cavalry regiments and three Ranger battalions under control of the Vietnamese 4th Armored Brigade. The attack lasted three days, and again was conducted without U.S. advisers or U.S. support once the troops were across the border. After two days of costly defeats, the enemy fled, and the South Vietnamese forces turned to evacuating large quantities of captured weapons and ammunition. The difficulties experienced in using armored cavalry assault vehicles to haul captured equipment prompted field commanders to request that in future operations ammunition caches be destroyed.

On 28 April the Vietnamese 2nd and 6th Armored Cavalry Regiments and Vietnamese Regional Forces attacked again into the Crow's Nest. American advisers were still not allowed on the ground in Cambodia, but for the first time U.S. support was used—command and control helicopters and gunships. The attack penetrated many enemy bases and diverted enemy attention from the larger attacks in III Corps Tactical Zone.

These early raids, a prelude to the major effort, helped to improve South Vietnamese procedures and techniques for use in more open warfare. They also afforded the free world forces a brief but eye-opening look at the massive size of the support facilities located across the border. Materiel and intelligence information confirmed in military minds the absolute necessity for large-scale operations into all the base areas. As a result of the attacks, the frustration built up over five years was vented, and success caused confidence and morale in the South Vietnamese Army to soar. Unfortunately, these operations also served to warn the Viet Cong and North Vietnamese that more attacks could be expected, and a hasty exodus of enemy units and headquarters to the west and north began.

After the enemy offensives of 1968, the tactics of free world forces underwent a change; from defensive, counterinsurgency tactics the allies began moving toward the offensive and toward the employment of more conventional tactics. The operations to secure the borders and clear the base areas in South Vietnam heralded this change. In the massive attacks into the border sanctuaries, which resembled exploitation or pursuit in conventional warfare, the change in tactics reached full course, and at all levels of command the difference was perceptible.

The Main Attack Into Cambodia

The major attack into Cambodia was a series of operations jointly planned and conducted by South Vietnamese and American units, directed at the highest levels, and involving the headquarters and forces of the South Vietnamese Army in the III and IV Corps zones and the U.S. II Field Force, Vietnam. When it began, Operation TOAN THANG 42, the Vietnamese portion, was probably the best planned South Vietnamese operation to that date. Weather and terrain were important considerations; it was recognized that any delay would invite considerable difficulties since the monsoon season would begin in late May. The weather in late April and early May would be good. The area chosen for the first attack was flat, with few natural obstacles to cross-country movement. The operation was planned by III Corps headquarters under conditions of great secrecy, and the participating Vietnamese units received only sketchy details until the plan was released on 27 April.[1] Early in the planning stage American advisers to South Vietnamese units acted as coordinators rather than advisers. Once the operation began, advisers were responsible for requesting and controlling American aid in the form of medical evacuation, close air support, and artillery fire. The advisers remained with their Vietnamese units up to thirty kilometers inside Cambodia; a presidential decree banned all U.S. ground participation beyond the thirty-kilometer line.

The operation was planned so that U.S. and South Vietnamese forces were separated by well-defined boundaries although they attacked simultaneously. This arrangement considerably simplified coordination and logistical planning and avoided possible confusion on the ground. The attack resembled a large double envelopment, with the South Vietnamese forces forming most of the western pincer and the American forces the center and the eastern pincer. *(Map 14)* Throughout the attack Vietnamese forces operated in combined arms task forces with infantry, artillery, and armored cavalry. The one exception was in the east in the airmobile assault of a South Vietnamese airborne brigade under U.S. control.

Operation TOAN THANG 42 began at 0710 on 29 April, less than forty-eight hours after the participating units were informed, when South Vietnamese task forces attacked to destroy enemy forces and supplies in Cambodia's Svay Rieng Province. The mission included opening and securing National Highway 1 to allow the evacuation of Vietnamese refugees and assisting the hard-pressed Cambodian Army to regain control of its territory.

All three task forces moving south and west in the Angel's Wing met the enemy during the first two days. The Viet Cong and North Vietnamese had anticipated the attack and fought a stubborn delaying action, hoping to evacuate supplies and equipment. Nonetheless, the first objectives were quickly achieved, and on 1 May the

[1] U.S. planning for Cambodia was also very secret; units were informed only twenty-four hours before the attack. However, the positioning of many artillery units and the sudden demand for Cambodian maps produced unfortunate security leaks. As in any big operation, the support units in the rear were informed first. The enemy definitely knew of the attack.

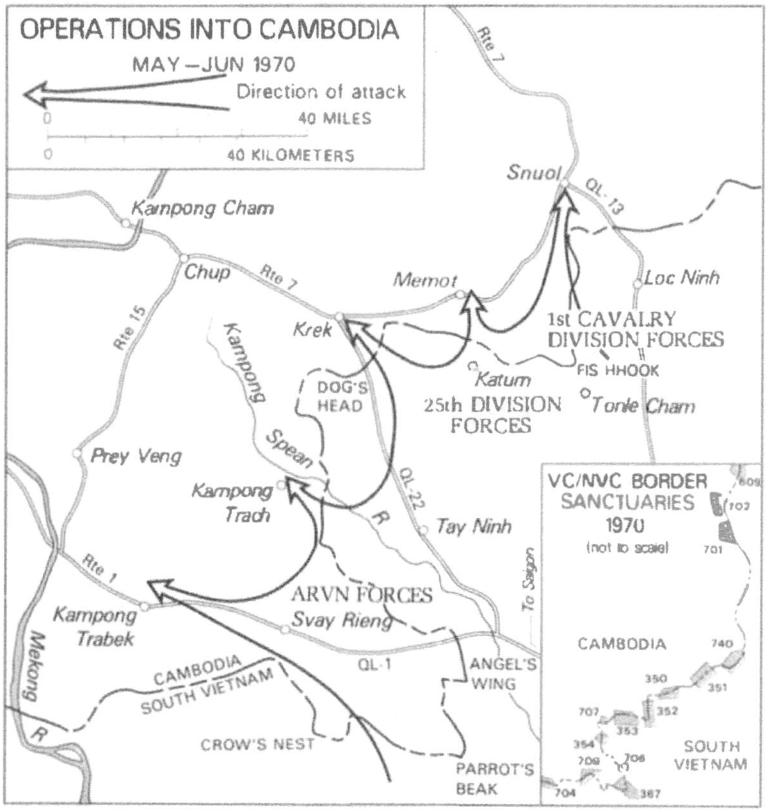

MAP 14

South Vietnamese swept west to the provincial capital of Svay Rieng, opening Route 1 to the east to Vietnam. The speed and success of this attack had in part been made possible by lessons learned in previous forays. The advance had been preceded by heavy air and artillery attacks on key areas inside Cambodia. Unlike the procedure in previous operations, assault units kept on attacking while follow-up units were responsible for the removal of captured supplies.

While the enemy's attention was riveted on the Angel's Wing region, the eastern pincer of the envelopment, TOANG THANG 43, was to attack under the command of Brigadier General Robert L. Shoemaker, assistant division commander of the 1st Cavalry Division (Air). His task force included the 3d Brigade of that division, the 2d Battalion, 34th Armor (—), the 2d Battalion, 47th Infantry (Mechanized), the South Vietnamese 3d Airborne Brigade, and the U.S. 11th Armored Cavalry.[2]

Following intensive air and artillery bombardment, the airborne brigade was to stage a helicopter assault into the area north of the Fishhook to seal off escape routes, while ground units attacked north. Air cavalry was to conduct screening operations as South Vietnamese cavalry screened the east flank in Vietnam. Task Force Shoemaker's mission was to locate and eliminate enemy forces and equipment. There was also a possibility that the Central Office for South Vietnam, the elusive enemy headquarters, would be found in the Fishhook and could be destroyed.

At 0600 on 1 May, U.S. artillery fire exploded on the proposed helicopter landing zones: the 1st Squadron, 9th Cavalry (Air), began aerial reconnaissance and was the first unit to find the enemy. Company C, 2d Battalion, 47th Infantry (Mechanized), closely followed on the west by the 2d Battalion, 34th Armor, led the attack of the 3d Brigade, 1st Cavalry Division (Airmobile) at 0945. To the east the 11th Armored Cavalry Regiment was hit at the border by elements of two enemy battalions. From their command vehicles with the lead tank platoon of the 2d Squadron, Colonel Donn A. Starry, the regimental commander, and Lieutenant Colonel Grail L. Brookshire, commanding the 2d Squadron, directed tactical air and artillery fire that immediately suppressed the enemy fire. The 11th Cavalry crossed the border at 1000.

The 2d Battalion, 47th Infantry (Mechanized), and the 2d Battalion, 34th Armor, proceeded north, unopposed, to secure landing zones to be used later in the day by the 3d Brigade, 9th Infantry Division.[3] The 2d and 3d Squadrons, 11th Cavalry, moved north with little opposition until late afternoon when Company H entered a clearing six kilometers inside Cambodia. Overhead, a scout helicopter from the regimental air cavalry troop discovered a large enemy force well entrenched on the edge of the clearing. The jungle suddenly erupted with enemy fire, and it quickly became evident that the enemy was on three sides of the 2d Squadron.

Colonel Starry immediately directed Lieutenant Colonel Bobby

---

[2] To bring the 11th Armored Cavalry together, the 3d Squadron moved more than 300 kilometers in two days and arrived witht every vehicle operational, just in time to join the attacking column.

[3] The 2d Battalion, 34th Armor, was withdrawn from Cambodia after only five days because of excessive tank breakdowns due to inadequate maintenance. The piecemeal employment of this unit in the entire war, together with its accompanying logistical problems, had contributed to this withdrawal.

F. Griffin, 3d Squadron commander, to attack the flank of the enemy defenses. The air cavalry hit the enemy's rear and his withdrawal routes, and at 1645 the enemy force, estimated at a battalion, broke and fled, leaving fifty-two dead. Two troopers of the 11th Cavalry were killed, the only American soldiers to die in Cambodia on 1 May.

By afternoon on 2 May free world forces were fighting in both wings of the envelopment. The South Vietnamese forces in the Parrot's Beak attacked south with two task forces from the III Corps Tactical Zone, while three task forces from IV Corps attacked north.[4] The object was to trap the enemy with elements of nine cavalry regiments. The 7th Squadron, 1st Cavalry (Air), sweeping ahead and to the flanks, had one troop credited with killing over 170 Viet Cong and North Vietnamese. The rapidly moving cavalry squadrons and U.S. air cover quickly broke the resistance of enemy troops and chased them into the guns of the other South Vietnamese task forces. The two South Vietnamese forces linked up early on the afternoon of 4 May. Over 400 of the enemy were killed: 1,146 individual weapons, 174 crew-served weapons, more than 140 tons of ammunition, and 45 tons of rice were captured.

In the eastern wing on 2 May, the 2d Battalion, 47th Infantry (Mechanized), cut Route 7 near Memot (Memut); the 2d Squadron, 11th Armored Cavalry, linked up with the South Vietnamese airborne forces. A search for supply caches met little resistance from enemy security forces.

Late on 3 May the 11th Cavalry was ordered to attack north forty kilometers to take the town of Snuol and its important road junction. Route 7, leading north to Snuol through large rubber plantations, was chosen as the axis of advance, and by early afternoon on the 4th, the lead tanks had broken out of the jungle and were on the ridge astride the highway. Once on the road the 2d Squadron, followed by the 3d, raced north at speeds up to sixty-five kilometers per hour and reached the first of three destroyed bridges by midafteroon. The cavalry secured the site, placed an armored vehicle launched bridge across the stream, and went on.[5]

With his regiment now strung out for almost sixty kilometers, Colonel Starry decided to consolidate south of the second stream

[4] On 2 May in IV Corps Tactical Zone, 250 ACAV's from five cavalry regiments lined up abreast at 25-meter intervals and, with infantry support, attacked on a 6-kilometer front. The breadth of the attack overwhelmed all resistance.

[5] While the 11th Armored Cavalry attacked north, Troop B, 1st Squadron, 9th Cavalry (Air), found an enemy logistics complex in the Fishhook that was so extensive it was named "the city" It yielded more than 1,500 weapons, millions of rounds of ammunition, and tons of supplies. This cache, the largest of the war, took several weeks to search and evacuate.

THE 2D SQUADRON, 11TH ARMORED CAVALRY, ENTERS SNUOL, CAMBODIA

crossing. Through the night, the 2d and 3d Squadrons closed on the lead elements, which were now reconnoitering the two remaining crossings. The 11th Armored Cavalry Regiment continued north on 5 May after Company H and Troop G laid another vehicle launched bridge at the second crossing site. The third crossing posed serious problems because it would require heavy bridging. A flying crane, the CH–54 helicopter, was requested to transport an M4T6 bridge to the site, but by midday when the 2d Squadron reached the third crossing site the crane pilots and the engineers had made little progress. Anxious not to lose the momentum of the attack, Colonel Starry set out on foot with the section sergeant and the bridge launching vehicle to find a place where the span could be used. After gingerly testing several places, they let down the bridge, tried it out with Troop G, and by 1300 the 2d and 3d Squadrons were again rolling north.

The 2d Squadron paused south of Snuol to bring up artillery, organize air support, and reconnoiter. Refugees reported that there were many North Vietnamese troops in the town and that the civilians had fled. Scouts from the regimental air cavalry troop had observed heavy antiaircraft fire all around the airstrip to the east of town. In midafternoon the 11th Cavalry surrounded the city, with the 2d Squadron on the east and the 3d Squadron on the west. As tanks and armored cavalry vehicles rumbled across the Snuol airstrip, they were hit by rocket propelled grenades and small arms fire, which ceased abruptly when the tanks replied with cannister. After a brisk fight, the antiaircraft guns were seized.

The 3d Squadron, meanwhile, moving through the rubber trees to encircle the town, triggered an ambush set to hit the 2d Squadron. Colonel Griffin placed artillery fire behind the enemy position, set up gunships to cover the right flank, and attacked with Troop I. As the 2d Squadron moved in from the southeast in a coordinated attack, an inexperienced gunship pilot fired rockets into the lead elements. This unfortunate incident caused the gunships to be withdrawn and opened one side of the trap as an escape route for the enemy. The two-squadron attack, however, routed the enemy troops, who fled in small groups in all directions. When the cavalry entered Snuol, the city was deserted.

Snuol was apparently the hub of an extensive logistical operation. On the following day, 6 May, the 2d Squadron discovered an improved road, large enough for trucks and carefully hidden under the jungle canopy. Along the road the cavalry found and destroyed an abandoned truck convoy laden with supplies. The cavalrymen also discovered in Snuol a fully equipped motor park, complete with grease racks and spare parts, and a large storage site containing 85-mm. tank gun ammunition.

While American units were attacking toward Snuol, South Vietnamese armored task forces in the southwest were expanding their area of operation to the north. Finding only disorganized enemy groups, the well-coordinated Vietnamese units quickly reached the Kampong Spean River and secured Kampong Trach. To the west of the town, an armor-heavy force overcame stiff resistance from three North Vietnamese battalions. The attacking armored units were closely followed by dismounted Rangers who eliminated the bypassed pockets of the enemy. This coordinated combined arms attack, supported by tactical air and artillery, demonstrated that the problems encountered during earlier South Vietnamese operations had been solved.

On 7 May President Nixon announced his satisfaction with the progress of the operations, and stated that U.S. troops would be withdrawn from Cambodia by 30 June. This announcement brought intensified search efforts, prompting additional attacks into Cambodia in the Dog's Head area and toward Krek by units of the 25th Infantry Division. The 1st Brigade moved on 14 May to come abreast of the 2d Brigade, which had been committed earlier in the Fishhook area. The move was led by the 2d Battalion, 22d Infantry (Mechanized), the 1st Battalion, 5th Infantry (Mechanized), and elements of the 3d Squadron, 4th Cavalry. These troops were to act as a blocking force north of the Kampong Spean for the South Vietnamese task forces that were now moving from Svay Rieng west to Kampong Trabek. Advancing in two columns on Route 1, and bypassing several small forces, a South Vietnamese task force covered over thirty kilometers to Kampong Trabek in slightly over two hours and linked up with South Vietnamese forces from IV Corps who were moving north toward Phnom Penh.

As the 18th Armored Cavalry Regiment was moving toward the linkup, the 15th Armored Cavalry Regiment collided with the 88th North Vietnamese Regiment, which was about to attack the 18th Armored Cavalry from the rear. Surprised, the enemy regiment tried to withdraw, but the fast-moving South Vietnamese force literally ran over it. A day-long, running battle left North Vietnamese Army resistance shattered, the enemy in flight, and the field covered with enemy casualties and abandoned weapons.

While the U.S. 25th Infantry Division held the center of the potential envelopment, a South Vietnamese task force moved north on 17 May to secure Route 15, halfway to the besieged town of Kampong Cham. Other South Vietnamese task forces spread out to secure the penetration, and Vietnamese district and province forces moved in to perform the detailed search and evacuation of captured material. On 9 May the 1st Squadron, 11th Armored Cavalry Regiment, which had been securing the line of communications to Tonle Cham in War Zone C, moved into Cambodia to search the rubber plantations at Memot. The cavalrymen discovered a motor park with twenty-one American-made 2½-ton trucks of World War II vintage that had been used in Korea (data plates were still on the vehicles), rebuilt in Japan, and sold as surplus. Once the batteries were replaced, the 1st Squadron had its own truck convoy to haul captured equipment and supplies back into South Vietnam.

As operations in and around the Fishhook continued it became evident that the free world forces had seriously underestimated the extent of enemy logistical bases in Cambodia. Consequently, the 11th Armored Cavalry Regiment was assigned two engineer land clearing companies and along with the Vietnamese airborne brigade began extensive Rome plow operations in the Fishhook. By late May the southern Fishhook, which had become the most hotly contested portion of TOAN THANG 43, contained two squadrons of the 11th Armored Cavalry Regiment. The monsoon had arrived, and movement became more difficult every day. As June wore on the enemy became more persistent, and small daily fights were a fact of life. Helicopters flying over the area habitually received ground fire. Road-mining incidents and ambushes increased. Nonetheless, the Rome plows cleared over 1,700 acres of jungle and destroyed more than 1,100 enemy structures.

In accordance with the presidential directive, plans were made to withdraw U.S. forces from Cambodia by 30 June. The 2d Squadron, 11th Armored Cavalry, moved along Highway 13 to Loc Ninh, while in the Fishhook the 1st and 3d Squadrons departed through Katum. The 3d Squadron was ordered to remove the bridge sections south of Snuol and to destroy the fire bases established along Route 7 near Memot, a job made more difficult as the monsoon inundated the low ground, leaving track vehicles virtually roadbound. With considerable difficulty the bridges were removed or destroyed by the 3d Squadron as it withdrew. Captain Ralph A. Mile's Troop L was the last U.S. armored unit to leave Cambodia.

South Vietnamese Army Attacks Continue

In late May, with the impending American withdrawal from Cambodia still to come and the monsoon rains increasing, the South Vietnamese forces, unhindered by the U.S. political decision, had continued attacking to complete the encirclement. The Chup rubber plantation near Kampong Cham was selected as the linkup point for the converging task forces. The city, a key provincial capital strategically located on the Mekong River, fifty kilometers northeast of Phnom Penh, was besieged by the 9th North Vietnamese Division, which had its headquarters in the plantation. The South Vietnamese objective was to attack the plantation from the south and east, thus eliminating enemy pressure on Kampong Cham and completely encircle the base areas.

At 0730 on 23 May, a South Vietnamese armored task force passed through the U.S. forces near Krek and ran head on into an entrenched rifle company from the 272d Viet Cong and North Vietnamese Regiment, intent on stopping any advance along Route 7. The enemy opened with rocket propelled grenade and small arms fire on the lead tanks of the 5th Armored Cavalry Regiment, which quickly formed on line and attacked. A short, fierce, close quarters fight left most of the enemy dead or captured. The action was so fast that units in the rear of the advancing column were unaware of the battle until they passed the enemy dead along the road. Another South Vietnamese armored task force moving north along Highway 15 on 25 May encountered an enemy battalion at the Chup plantation. The 15th and 18th Armored Cavalry Regiments hit the enemy positions from two sides, completely disorganizing the resistance. Fire from the advancing tanks and armored cavalry assault vehicles left over 110 North Vietnamese dead. Three days later in the southwestern part of the Chup plantation, another enemy battalion was defeated in a six-hour action. The task forces joined on 29 May, breaking the seige of Kampong Cham and surrounding the enemy base areas to the south.

In mid-June, when the 9th North Vietnamese Division re-entered the Chup plantation, again threatening Kampong Cham and Prey Veng, South Vietnamese units began a new drive to clear the plantation and destroy the enemy division. For six days an armor-heavy task force chased the enemy through the rubber plantation and south along Route 15. Finally, the task force was attacked by the 271st Viet Cong and North Vietnamese Regiment, which planned to cut the road and isolate the task force. The next three days saw some of the heaviest fighting of TOAN THANG 42. The enemy was well positioned, but repeated attacks by the 15th and 18th Armored Cavalry Regiments prevented his control of the road. Attacks by tanks and armored cavalry assault vehicles with attached infantry and Rangers finally routed the North Vietnamese, and by 29 June the fighting had ended. The final phase of the operation, 1–22 July, involved road security missions and search operations. No significant fighting occurred, and all Military Region 3 units had left Cambodia by 22 July 1970.[6]

The South Vietnamese forces of Military Region 3 periodically returned to Cambodia during the next eighteen months. Most of their operations were hit and run, and had limited objectives and minor successes. One operation, TOAN THANG 01–71, started with high hopes on 4 February 1971 and ended in disaster four months later. Little was officially reported about this operation since world attention was dramatically focused on LAM SON 719 to the north in Laos, and no U.S. ground forces or advisers could be used in Cambodia. Near the end of May a South Vietnamese task force was cut off in Cambodia south of Snuol on Route 13. Although the commander had received intelligence reports from South Vietnamese and U.S. sources, including visual aerial reconnaissance from the 3d Squadron, 17th Cavalry (Air), he failed to guard against a growing enemy threat.[7]

The 3d Armored Brigade was ordered north to link up with the isolated task force on Route 13. After a misunderstanding of orders, during which the task force at first refused to attempt to withdraw,

---

[6] Corps tactical zones were designated military regions on 1 July 1970.

[7] Air cavalry support was provided by the largest grouping of air cavalry ever assembled under one commander. The 1st Squadron, 9th Cavalry (Air), 3d Squadron, 17th Cavalry (Air), an aerial rocket artillery battalion, and some separate air cavalry troops were formed into the 1st Brigade, 9th Cavalry (Air). This was the first air cavalry combat brigade ever formed and used in combat.

the task force troops attacked south toward the armored brigade, but intense rocket, small arms, and machine gun fire quickly disorganized the attack. Many infantrymen ran from exploding or disabled vehicles; others, trying to hide from the deadly fire, climbed under or onto the vehicles. Some soldiers already on the vehicles crawled inside for cover, while more and more attempted to mount the moving vehicles to escape. As the column continued down the road it became a rout.

After two days of massive artillery, air cavalry, and tactical air strikes, the armor brigade finally accomplished the linkup. The task force passed through the brigade with each tracked vehicle carrying thirty to thirty-five infantrymen. Both sides suffered heavily, but for the South Vietnamese forces command and control again emerged as a serious problem. What had begun as an orderly withdrawal, turned into a rout. The collapse of command under stress was to plague the South Vietnamese forces to the end of the war.

Secondary Attacks Across the Border

In examining the final results of the expeditions into Cambodia, it is well to note that two separate series of South Vietnamese operations supported the TOAN THANG attacks. The first, Operation CUU LONG I–III, from IV Corps Tactical Zone, lasted from 9 May till 30 June 1970, and involved five armored cavalry regiments as well as infantry, Rangers, elements of the Vietnamese Navy, and units of the Regional Forces and Popular Forces. The operational area was more than ninety kilometers wide and extended north to Phnom Penh. The object was to secure the Mekong River as far north as the Cambodian capital so that the Vietnamese refugees gathered in the city could be evacuated to South Vietnam. The operation indirectly supported III Corps Tactical Zone forces involved in Operation TOAN THANG 42.

CUU LONG I began with an assault on locations along the Mekong by aircraft of the U.S. 164th Aviation Group which formed the largest air armada ever assembled in IV Corps Tactical Zone for a single operation. By 13 May linkup had been accomplished with III Corps forces, and the Mekong River was secure from the border to the capital of Cambodia. Five hundred of the enemy were killed. Forty ships passed safely up the Mekong to Phnom Penh, where they evacuated over 12,000 Vietnamese civilians. Eventually, more than 40,000 Vietnamese were evacuated through this safe corridor.

Operations CUU LONG II and III, from 17 May to 30 June, were directed at enemy forces and base camps in southeastern Cambodia. They were designed to assist the Cambodians in constructing bases and reestablishing local government. In both operations cavalry units traveled rapidly for more than fifty kilometers to relieve besieged Cambodian garrisons, and then turned their attention to searching for supplies. One cache discovered by Troop D, 3d Squadron, 5th Cavalry, yielded millions in National Liberation Front money printed for use in South Vietnam after *Tet* 1968.

The other series of operations into Cambodia originated in II Corps Tactical Zone and was designed to support TOAN THANG 42 by drawing enemy units north and cutting the enemy logistical lifeline north of the main battle. Operations BINH TAY I–IV were conducted from Kontum in the north to Ban Me Thuot in the south and were controlled and executed by the South Vietnamese Army.

The first three phases of BINH TAY were directed against Base Areas 701, 702, and 740, long utilized to support Viet Cong and North Vietnamese units operating in the central highlands of the II Corps area. There was little activity by armored units during these operations; South Vietnamese commanders preferred to use their armor for security and transportation. Enemy resistance was light and poorly organized. These first phases, although successful, showed clearly that the South Vietnamese commanders in II Corps Tactical Zone did not fully appreciate the possibilities for maneuver and firepower that armored units possessed. The II Corps cavalry regiments were not given the freedom of action afforded similar units in the III and IV Corps areas.

BINH TAY IV, conducted from 24 to 26 June, was the final II Corps operation in Cambodia. It included the largest aggregation of armored forces in the II Corps zone and, unlike other BINH TAY operations, was not directed toward destruction of enemy forces or bases but toward the evacuation of Cambodian and Vietnamese refugees. The armor spearhead, catching the enemy units off guard, moved swiftly into Cambodia on 24 June and set up defensive positions along the withdrawal routes. When the operation ended on 26 June 1970, over 8,500 Cambodians, more than 3,800 of them military, and over 200 vehicles and much equipment had been removed from the danger of control by the Viet Cong and North Vietnamese.

Cambodia in Perspective

By the end of June free world forces in Cambodia had captured or destroyed almost ten thousand tons of materiel and food. In terms of enemy needs this amount was enough rice to feed more than 25,000 troops a full ration for an entire year; individual weapons to equip 55 full-strength battalions; crew-served weapons to equip 33 full battalions; and mortar, rocket, and recoilless rifle ammunition for more than 9,000 average attacks against free world units. In all, 11,362 enemy soldiers were killed and over 2,000 captured.

These statistics are impressive, and without a doubt the Cambodian expeditions had crippled Viet Cong and North Vietnamese operations, but the most important results cannot be measured in tangibles alone. The armored-led attacks into Cambodia by units from Military Region 4 had been well planned, well coordinated, and well carried out. They were generally conducted without the massive U.S. ground support typical of operations by units from Military Region 3, yet they severely hurt the enemy. The South Vietnamese, their morale high, returned to resume pacification of the delta, a goal which had suddenly come much closer to realization.

In Military Region 3 the results of operations TOAN THANG 42 and 43 were alo impressive, and had a great pyschological and material effect on the enemy. Even more important, South Vietnamese forces had operated over great distances for long periods without direct American assistance and often without advisers. This fact provided a great boost to South Vietnamese morale and improved fighting ability. The Vietnamese forces had temporarily strengthened the position of the Cambodian government and brought some measure of order to its border provinces.

On the other side of the ledger, the results of the last expedition from Military Region 3 revealed the continued existence of command and control problems among South Vietnamese commanders. To overcome timidity and lack of coordination at high command levels, would, in the final analysis, be more important than material gains.

The lack of understanding of armored operations exhibited in Military Region 2 did not bode well for the future, although eventually new commanders there would begin the process of correction. The boost given to the Cambodian government and its army was only temporary, for the Viet Cong and North Vietnamese forces returned and quickly took complete control of all the border areas.

The most important effect of the operations in Cambodia must be looked for within South Vietnam: here the attacks bought time for strengthening the Vietnamese forces and for the United States to continue its withdrawals.[8] For the next fourteen months there

---

[8]In April 1970 the first battalion-size tank, cavalry, and mechanized units left Vietnam: the 1st Battalion, 69th Armor; 1st Squadron, 4th Cavalry; 2d Battalion, 2d Infantry (Mechanized); and 1st Battalion, 16th Infantry (Mechanized). These last three units were the armored strength of the 1st Infantry Division.

were almost no Viet Cong and North Vietnamese operations in South Vietnam. The Cambodian operations greatly increased the confidence of Vietnamese armored forces in their ability to wage a successful and prolonged campaign. It was the most convincing evidence since *Tet* 1968 of the improvement of Vietnamese armored forces. The high morale of the South Vietnamese forces convinced American advisers that the Vietnamese were well on their way to being able to fight the war on their own.

Maintenance and Supply

One unusual feature of American operations in Cambodia was the American policy that all vehicles and equipment, no matter how badly damaged—even if beyond repair—had to be evacuated to Vietnam. While this ruling was made at high levels for political, intelligence, and propaganda purposes, it calls attention to a major problem that confronted armored units throughout the war.

In an armored unit, the soldier is as dependent on his armored vehicle as the vehicle is on him. In few other combat units in the Army are maintenance and the supporting supply system so critical. Most armored units found the U.S. Army supply and maintenance system in Vietnam to be less than satisfactory at every level. The deficiencies in the system were basic. As the war expanded and mobility became more and more important the faults in the system became more obvious.

Two faults were apparent in the early years, and they eventually exposed a third. The first was the lack of general support maintenance—heavy repair facilities in the major areas where armored vehicles were used. This lack was the result of the decisions of 1965 and 1966 to build up combat troops at the expense of the logistical base. Although it was an expedient meant for a short time, the decision was never really altered. Even more unfortunate was the fact that in many cases the few support units that were available were centralized in areas far from the combat units. The obvious solution to this problem, the use of teams authorized to make major repairs at a unit's location, however popular with units was not popular with logisticians. Thus, combat units were frequently forced to send damaged vehicles great distances for repair. In Military Region 3, vehicles were almost always sent back to the Long Binh–Saigon–Cu Chi area, a distance of ninety or more kilometers from the border and base areas where the fighting was. The resulting loss in combat power and the drain on the meager evacuation resources of the combat units was a severe hardship.

In an attempt to solve this problem, Colonel Starry had forced repair teams forward to squadron and troop level in the 11th Armored Cavalry Regiment even before the invasion of Cambodia. To help with the critical problem of evacuating materiel from Cambodia the 11th Cavalry borrowed six M88 tank recovery vehicles from the depot at Long Binh. Organized into recovery platoons operating with the 3d Squadron, these vehicles were invaluable to the regiment's recovery operations.

The second problem, the tendency of logistical units to stick to base camps, was evident early in the war and continued to the end. Logistical units, particularly supply and maintenance elements, were unprepared psychologically and in practice to live in the field close to the units they supported. Although Army doctrine stressed that this support should be provided in forward areas, the practice was to centralize support facilities in built-up, well-developed, permanent base camps, similar to installations in the United States. In Military Regions 2 and 3, this practice placed support facilities as close to the coast as possible, often more than 100 kilometers from the fighting units, and accessible only by means of tenuous supply and evacuation routes. While this placement was easier for the supply and maintenance units, it was a hardship for the combat units.

The most critical problem was the unsatisfactory performance of the area support system under combat conditions. It is amazing that the system was expected to work in a war of movement, in which armored units traveled great distances in short periods of time. In Vietnam, it was not the answer for armored units, particularly armored cavalry regiments. According to Lieutenant General Joseph A. M. Heiser, Jr., former commander of the 1st Logistical Command, Vietnam, the 11th Armored Cavalry Regiment "obtained its maintenance support from the 1st Logistical Command on an area basis. As elements of the regiment relocated, the nearest 1st Logical Command unit provided service. This method of support proved unsatisfactory because of the 11th ACR's high and fluctuating maintenance demands. In the future such organizations should be assigned an organic maintenance unit."

While the problem was apparent in the 11th Armored Cavalry, it existed also for the 1st Brigade, 5th Infantry Division (Mechanized), in northern Military Region I. Although the mechanized brigade operated under control of the 3d U.S. Marine Division, the marines were responsible for supplying only rations and fuel. The 1st Brigade had its own organic supply and maintenance support but relied for wholesale level supply on distant Army support units. Several unanticipated maintenance difficulties developed as a result of this extended logistical link. Operations in the sandy soil of the northern coastal areas caused such excessive track and sprocket wear that spare parts were frequently inadequate. To cope with the problem, special brigade convoys were sent directly to the depots in an effort to shorten the delivery time of parts. Nicknamed the Red Ball Express, these convoys eventually eased many of the brigade maintenance problems.

These were not isolated examples, for combined arms operations stressed cross-attachment at battalion level, with the units often operating over great distances for long periods of time. The 2d Battalion, 34th Armor, in Military Region 3 had two companies detached for almost its entire time in the war—four years. Maintenance support of these two companies remained the responsibility of the battalion. Since one company was in northern Military Region 1, the situation became ludicrous when the battalion executive officer had to search the Saigon area for parts and then airlift them to the Hue area, 750 kilometers north. The other company, in Military Region 3, was often split among three locations, yet the battalion had to find and support them daily, even though the company was attached to a different division. Since the battalion had a limited resupply and maintenance system, this situation was completely unsatisfactory.

In Military Region 2, elements of the 1st Battalion, 69th Armor, and 1st Battalion, 50th Infantry (Mechanized) were often attached to the 1st Air Cavalry Division, yet the division had no means of repairing armored vehicles. With the distant parent battalion still responsible for maintenance and logistical support, the tank and mechanized companies frequently found themselves sorely pressed for supplies and replacement parts. The 1st Cavalry Division was able to repair equipment common to both the companies and the division; however, if the equipment was not common, long resupply delays were normal. In an attempt to partially solve the problem, Company A, 1st Battalion, 69th Armor, used its organic vehicles to obtain repair parts directly from the Qui Nhon support command.

Examples of the failure of the rigidly structured area support system to sustain adequately a constantly changing troop concentration are almost endless. Support units requisitioned parts over a 9,600-mile supply line, with attendant delays. There were never enough spare parts on hand to repair the armored vehicles in any given area. When parts were ordered they often arrived after the units had moved to a different support area, and the requisitioning process had started again.

Many demands that could have been met by depots in Vietnam were not met because the centralized inventory system broke down. Depots had spare parts, some of them important items, of which they were not even aware. Thus was born the system of searching for parts that the combat units called scrounging. The scrounger, or expediter, was an individual or a team from the combat unit sent to the major spare parts depot, usually in the Saigon area, to walk through the storage areas in an attempt to locate spare parts. When an item was found, it had to be formally released by supply control officials, often over the protests of supply personnel who were positive they did not have the item—even when it was physically pointed out. This fact was recognized by the Department of the Army late in the conflict when projects such as Stop/See, Count, Condition, and Clean were started in an attempt to verify inventories. One inventory team was sent to Okinawa to open twenty-six acres of shipping containers for which no inventory existed.

In combat units, inexperienced crew members, supervisors (officers and noncommissioned officers), and maintenance personnel contributed to the problem. Parts were often requested and replaced unnecessarily. Such instances became more apparent late in the war as a greater number of untrained people were assigned to unit maintenance operations.[9] Compounding this problem was the sometimes improper management of the battalion parts system that led to inadequate records and failure to order parts. In 1969 and 1970 the 11th Armored Cavalry was able to reduce its prescribed load lists by about 75 percent, with a dramatic increase in operational readiness. The supply system at the unit level was glutted with too much unneeded gear. Nonetheless, Colonel Starry noted that his regiment was obliged to live off its battle losses by cannibalizing disabled vehicles: the supply system provided only half the regiment's needs, cannibalization the rest.

The critical problem continued to be with the area support system. Although commented upon in the 1967 report evaluating mechanized and armor combat operations in Vietnam, a maintenance support unit dedicated primarily to the 11th Armored Cavalry Regiment was not created until 1970, and then only after the regimental commander had convinced officers at higher levels that such a measure was necessary.

Evacuation of damaged vehicles was another problem that

[9] Many commanders felt that a periodic standdown for maintenance was necessary. Others felt that this policy resulted in the delay of daily maintenance tasks until the scheduled standdown. The latter group, which favored preventive maintenance, recommended constant unit maintenance in the field with the help of teams from maintenance support units.

M88 HEAVY RECOVERY VEHICLE LOADS DAMAGED APC, JANUARY 1971. *This versatile vehicle was the workhorse of the armor recovery fleet.*

plagued combat units. Before the Vietnam War the standard practice had been to leave damaged vehicles at collecting points on the main supply route for supporting units to dispose of. But in Vietnam no provision for such evacuation was made, and the responsibility therefore fell entirely upon the combat units. In Cambodia, for example, combat units evacuated all vehicles to Vietnam no matter how badly damaged. The 11th Armored Cavalry Regiment eventually devoted more than one-third of its combat strength in Cambodia to this task.

Recovery of damaged machines at the small unit level required considerable ingenuity. Often the vehicles designed for recovery were inadequate, as in the case of the M578, or were in short supply, like the M88. No unit ever had these items in the numbers required, and there were never enough spare parts to repair them on the spot. Because recovery vehicles were frequently out of action for extended periods, awaiting parts, heavy reliance had to be placed on the inventiveness of the small unit leader. In many cases the performance of these leaders was brillant. Such recovery devices as the push-bar, log extraction, "daisy chain," and block and tackle were field expedients.

Many of the maintenance problems cited, particularly the attitude of the snpport units, exist today and would create the same difficulties in a war fought in Europe as they did in Vietnam. In view of the renewed emphasis on mobile warfare and the heavy odds that armored forces must face, a much more responsive supply and maintenance system is a necessity. Forward location of maintenance units, forward support, mobile repair teams, and quick resupply from accurate inventories must become as routine for combat service support units as the use of combined arms for armored units.

Lam Son 719

After U.S. units withdrew from Cambodia in June of 1970, the face of the war in Vietnam changed significantly. The remainder of the year was a time of small and infrequent enemy infantry attacks, fire attacks, and chance engagements. American forces directed their efforts toward strengthening the South Vietnamese forces and pacification of the South Vietnamese people. Mainly because of the Cambodian incursions and the resulting disruption of the enemy's supply and training bases, both causes advanced rapidly.

American forces were relocated in bases farther from the border, and the South Vietnamese Army assumed responsibility for the security of the border.[10] For the first time in many years, the South Vietnamese had to shoulder the larger share of combat operations—a dramatic change. South Vietnamese forces moved toward self-sufficiency and achieved considerable success. Regional Forces and Popular Forces took over many of those defensive operations that had long tied down the Vietnamese Army. And as U.S. troops were withdrawn from Vietnam, South Vietnamese units began large-scale operations on their own.

By late 1971, after extensive destruction of enemy supplies during the Cambodian incursions, enemy logistical and troop movements along the Laotian trails in the north increased dramatically. This fact and the impending withdrawal of U.S. air support prompted the South Vietnamese Army to attack into Laos and strike the enemy trail network at a junction near Tchepone. *(Map 15)* The South Vietnamese planned to commit two reinforced army divisions and their Marine division to this operation, LAM SON 719, commencing early in 1971. The planners considered this attack the last chance for cross-border operations using U.S. air support. They also believed that the operation, if successful, could

10 The transition after Cambodia was difficult for armored units since they had been used to responding to enemy action with massive firepower. In Vietnam, the population density away from the jungles prevented this and fire control had to be again closely supervised.

prevent a major enemy offensive for at least another year and take some pressure off the Cambodian Army to the south.

LAM SON 719 demonstrated what can happen when a large operation is insufficiently coordinated: conflicting orders were issued, the limited amount of armor was misused, unit leadership broke down, and the strength of the enemy was either overlooked or disregarded. That the North Vietnamese knew of the attack beforehand was evident in their placement of artillery, mortars, and antiaircraft weapons in the area of operations chosen by the South Vietnamese. Enemy troop buildups north of the Demilitarized Zone were noted as well as an increase in the movement of supplies along the trails.

Although American ground forces supported LAM SON 719, they were required to remain in South Vietnam. A task force, part of Operation DEWEY CANYON II, consisting of elements of the 1st Battalion, 61st Infantry (Mechanized),[11] the 1st Battalion, 77th Armor, the 3d Squadron, 5th Cavalry, and Troop A, 4th Squadron, 12th Cavalry, had the mission of establishing logistical bases, keeping Route QL-9 open to the Laotian border, and covering the withdrawal of the South Vietnamese.

At 0400 on 29 January the task force left Quang Tri City along National Highway 9 and by nightfall rolled into Fire Support Base Vandergrift. After a short halt Troop A, 3d Squadron, 5th Cavalry, commanded by Captain Thomas Stewart, and two engineer companies led out on foot at midnight on 29 January. The vehicles were left to move with the main body since Route 9 was known to be in a poor state of repair. A bulldozer led the column with headlights blazing.[12] Whenever an obstacle such as a damaged bridge was encountered, a force of two to six cavalrymen and engineers would stop to make repairs while the rest of the team continued. The cavalry troop, joined by its vehicles, arrived at Khe Sanh at 1400 on 1 February, with National Highway 9 opened behind it from Fire Support Base Vandergrift. The next day the road was opened all the way to the border by the 1st Squadron, 1st Cavalry (–).

As a supplement to this route, the remainder of the 3d Squadron, 5th Cavalry, and elements of the 7th Engineer Battalion con-

---

11 Brigadier General John G. Hill, Jr., 1st Brigade commander, had reorganized the mechanized infantry and equipped each platoon with six APC's, each manned by a six-man squad. This organization increased the firepower of each platoon and decreased the load on each vehicle.

12 Originally, an M551 Sheridan had started to lead the column with its infrared light, but the Sheridan and an accompanying mortar carrier were unable to negotiate the rugged terrain and turned back.

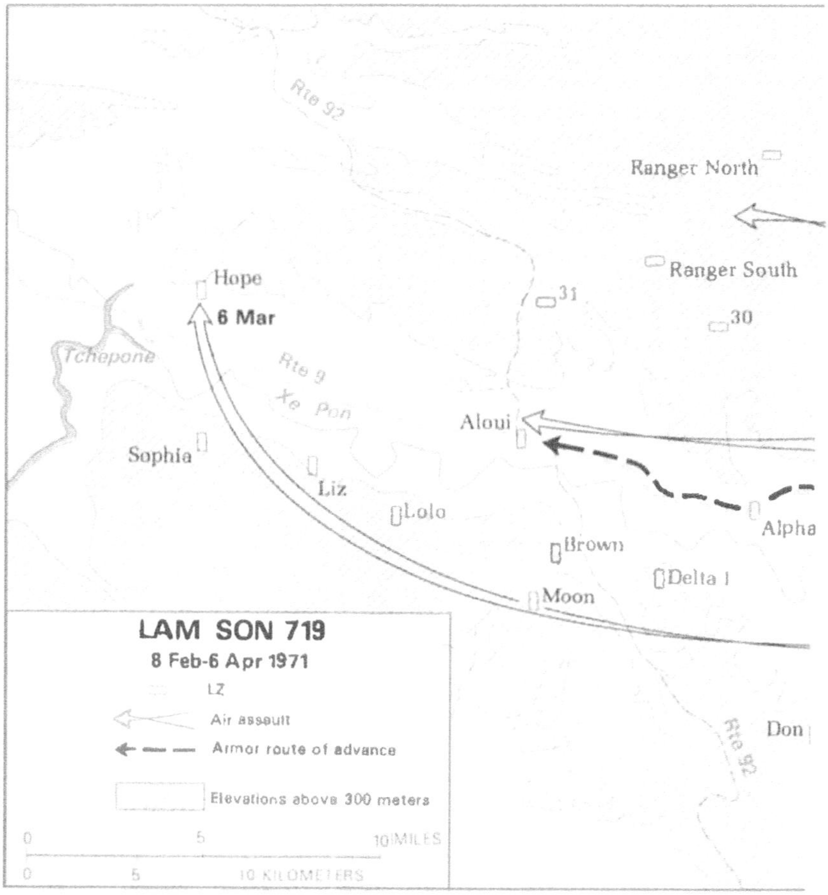

Rte 92
Ranger North
Ranger South
Hope
6 Mar
31
30
Tchepone
Rte 9
Xe Pon
Aloui
Sophia
Liz
Lolo
Alpha
Brown
Delta 1
Moon
Don
Rte 92
LAM SON 719
8 Feb-6 Apr 1971
LZ
Air assault
Armor route of advance
Elevations above 300 meters
0 5 10 MILES
0 5 10 KILOMETERS

MAP 15

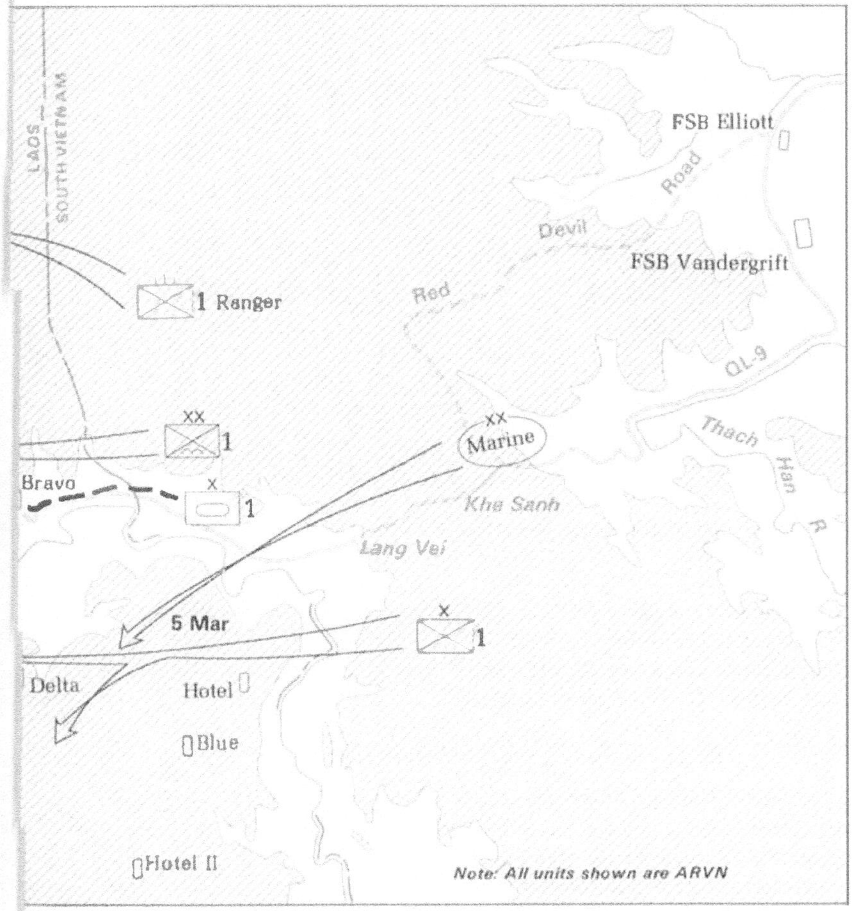

LAOS
SOUTH VIETNAM
FSB Elliott
Road
Devil
Red
FSB Vandergrift
1 Ranger
QL-9
XX
1
XX
Marine
Thach
Han
R
Bravo
X
1
Khe Sanh
Lang Vei
5 Mar
X
1
Delta
Hotel
Blue
Hotel II
Note: All units shown are ARVN

RED DEVIL ROAD, *an engineering feat that opened enemy areas never before penetrated.*

structed a secondary road, known as Red Devil Road and roughly parallel to Route 9, from Fire Support Base Elliott to Khe Sanh. The 3d Squadron, 5th Cavalry, continued operations north of Khe Sanh along Red Devil Road until 7 April.

The South Vietnamese Army Attack

In LAM SON 719, the Vietnamese hoped to disrupt Viet Cong and North Vietnamese supply lines by a combination of airmobile and armor ground attacks on three axes westward into Laos. The main attack was to be conducted along National Highway 9 to Aloui by the airborne division and the 1st Armored Brigade, which would then continue west on order. The South Vietnamese 1st Infantry Division, in a series of battalion-size airmobile assaults, was to establish fire bases on the high ground south of Route 9 to secure the south flank. The South Vietnamese 1st Ranger Group was to conduct airmobile assaults to establish blocking positions and secure the north flank. The Vietnamese Marine division was the I Corps reserve at Khe Sanh. The U.S. 2d Squadron, 17th Cav-

ACAV's of South Vietnamese 1st Armored Brigade on Route 9 in Laos, 1971

alry, was to locate and destroy antiaircraft weapons, find enemy concentrations, and carry out reconnaissance and security missions, which included the rescue of air crews downed in Laos. The squadron was permitted to go into Laos only one hour before the first airmobile assaults. This constraint precluded early reconnaissance of North Vietnamese antiaircraft positions, and in the beginning limited the air cavalry to screening the landing zones just before the assaults.

The 1st Armor Brigade, with two South Vietnamese airborne battalions and the 11th and 17th Cavalry Regiments, which had fewer than seventeen M41 tanks, crossed the border at 1000 on 8 February and moved nine kilometers west along National Highway 9. Intelligence reports had indicated that the terrain along Route 9 in Laos was favorable for armored vehicles. In reality, Route 9 was a neglected forty-year-old, single-lane road, with high shoulders on both sides and no maneuver room. Moreover, as the units moved forward they discovered the entire area was filled with huge bomb craters, undetected earlier because of dense grass and bamboo. Armored vehicles were therefore restricted to the road.

With armored units moving west on Route 9, the airborne division and the 1st Infantry Division made an assault into landing zones north and south of Route 9. One Ranger battalion came down near Landing Zone Ranger South. As the first troops arrived the air cavalry moved out to reconnoiter the front and flanks, seeking landing areas and destroying antiaircraft positions. But the demand for gunships became heavy as units on the ground encountered North Vietnamese Army forces. In the air cavalry, emphasis shifted to locating enemy troop concentrations and indirect fire weapons that posed an immediate threat to South Vietnamese forces. Thus, long-range reconnaissance was sacrificed for fire support.

The air cavalry screened the 1st Armor Brigade's advance along Route 9 all the way to Aloui, which the brigade reached in the afternoon of 10 February.[13] Within three days Vietnamese airmobile forces on the ridgelines to the north and south had moved abreast of Aloui. Since the airborne division was unable to secure Route 9, the 1st Armor Brigade as well as other ground forces had to be resupplied by air for the duration of LAM SON 719.

Enemy reaction to LAM SON 719 was swift and violent. The North Vietnamese had elements of three infantry regiments as well as an artillery regiment and a tank battalion in the area, and quickly brought in eight more infantry regiments and part of a tank regiment. The north flank of the South Vietnamese attack soon came under heavy assault. The Ranger battalion at Landing Zone Ranger North was attacked on 20 February, and elements of the battalion withdraw to Landing Zone Ranger South the next day. In the following days both Ranger South and Landing Zone 31 came under increasing pressure until, on 25 February, the Rangers were evacuated from Ranger South.

As the South Vietnamese command debated whether to continue the drive west, pressure on Landing Zone 31 developed into a coordinated enemy tank-infantry attack with supporting fire from artillery and rockets. Command confusion added to the problems of the Vietnamese forces when conflicting orders from the airborne division and from I Corps headquarters delayed relief of the landing zone by the armored brigade. On 18 February I Corps ordered the 17th Armored Cavalry (–) north from Aloui to reinforce Landing Zone 31. At the same time the airborne division ordered it to stop south of the landing zone and wait to see if the site was overrun. Neither headquarters was on the scene. As a result of the confusion, the 17th Armored Cavalry, with tanks from the 11th Armored Cavalry, arrived at Landing Zone 31 on 19 February after some airborne elements had been pushed back.

In the first battle between North Vietnamese and South Viet-

---

13 During the advance Troop C, 2d Squadron, 17th Cavalry (Air), discovered a modern enemy oil pipeline west of Aloui and destroyed several sections by Cobra fire. Eventually, this pipeline was rebuilt by the enemy and extended many miles south into Military Region 3.

namese tanks, Sergeant Nguyen Xuan Mai, a tank commander in the 1st Squadron, 11th Armored Cavalry Regiment, destroyed a North Vietnamese T54 tank.[14] The South Vietnamese forces retook a portion of the landing zone by the end of the day. Twenty-two enemy tanks—six T54's and sixteen PT76's—were destroyed, with none of the South Vietnamese M41's lost. Direct and indirect fire continued to pound the airborne troops, and, finally, after six days, the enemy overran the entire landing zone. The 17th Armored Cavalry Regiment and one airborne battalion were pushed to the south.

After Landing Zone 31 was lost, all airborne elements were withdrawn and the 17th Armored Cavalry was isolated southeast of the site. Enemy pressure on the cavalry remained heavy. Attacked at noon on 27 February the cavalry, supported by tactical air and cavalry helicopter gunships, reported destroying fifteen tanks—twelve PT76's and three T54's—and losing three armored cavalry assault vehicles. Later, on 1 March, still southeast of Landing Zone 31, the cavalry was attacked again. In this battle, which lasted throughout the night, the cavalry was supported by South Vietnamese artillery, U.S. tactical air strikes, and cavalry gunships. Fifteen enemy tanks were destroyed; the cavalry lost six armored cavalry assault vehicles.

Despite recommendations from the American adviser of the 1st Armor Brigade and the acting adviser of the division, the commander of the airborne division failed either to support the 17th Armored Cavalry or to withdraw it. On 3 March, after the cavalry was surrounded on three sides by enemy armor and its ronte of withdrawal was blocked by direct tank gunfire, the South Vietnamese Chief of Armor, with the approval of the I Corps commander, intervened by radio. He obtained air support from I Corps and ordered the 17th Cavalry south to more defensible ground. From there, the cavalry subsequently fought a delaying action and rejoined the 1st Brigade at Aloui.

Air Cavalry and Tanks

Fortunately for Operation LAM SON 719, the confusion on the ground did not extend to the air cavalry. The performance of the air cavalry remains one of the outstanding achievements of the operation, particularly since it operated in the most hostile air environment of the war. All air cavalry in Laos was controlled by

14 The South Vietnamese enemy identification books pictured only the Soviet PT76 and T54 tanks. Thus, all tanks not PT76's were identified as T54's. In this case the T54 was correctly identified by a photograph of the T54 tank.

the U.S. 2d Squadron, 17th Air Cavalry, which reported directly to the U.S. XXIV Corps. In addition, the cavalry had operational control of the reconnaissance company of the South Vietnamese 1st Infantry Division. Called the Black Panthers, or Hac Bao, the unit was an elite, 300-man company, cross-trained and organized into aerorifle platoons, and used for ground operations in Laos.

The greatest threat to air cavalry was fire from .51-caliber machine guns, which the North Vietnamese Army employed in large numbers, locating them in mutually supporting positions. The OH–6A scout helicopter was too vulnerable to heavy fire from these guns to operate as part of the reconnaissance team. Instead, groups of two to six AH–1G Cobras and one command and control aircraft were formed, with scout pilots as front seat gunners in the Cobras. Although not designed as a scout ship, the Cobra did well in the reconnaissance role. Its weapons could immediately engage the enemy and it was powerful enough to make runs at high speed through hostile areas without taking unacceptable risks.

When the squadron encountered tanks for the first time, high-explosive antitank (HEAT) rockets were not available, and it used whatever ordnance was on board. The Cobra gunships opened fire at maximum range, using 2.75-inch flechette rockets to eliminate enemy troops riding on the outside of the tank and to force the crew to close the hatches. As the gun run continued, high-explosive and white phosphorus rockets and 20-mm. cannon fire were used against the tank itself.

Eventually HEAT rockets became available, but they were not always effective. Although these rockets were capable of penetrating armorplate, they could do so only in direct hits. Engagements therefore had to take place at ranges of 900 to 1,200 meters, distances that exposed the gunship to the tank's heavy machine gun and to supporting infantry weapons. Between 8 February and 24 March, air cavalry teams sighted 66 tanks, destroyed 6, and immobilized 8. Most of the tanks, however, were turned over to fixed-wing aircraft, which could attack with heavier ordnance.

The Withdrawal

After the 17th Armored Cavalry withdrew from Landing Zone 31 and returned, the 1st Armor Brigade task force continued to occupy bases near Aloui. Again because of conflicting orders from the airborne division and I Corps headquarters, the brigade did not move farther west and therefore became a target for intense enemy fire; losses in men and equipment mounted. Eventually a point was reached when the 1st Armor Brigade could not, if it had been ordered, move west of Aloui. As a result, the 1st Infantry Division was ordered to seize Tchepone, and did so on 6 March with an airmobile assault into Landing Zone Hope.

By early March enemy forces in the LAM SON 719 area had increased to five divisions: 12 infantry regiments, 2 tank battalions, an artillery regiment, and at least 19 antiaircraft battalions. After encountering enemy armored vehicles at LandingZone 31, South Vietnamese planners had realized that North Vietnamese armor was present in strength, and the 1st Armor Brigade was strengthened with additional units as they became available. The reinforcement was so piecemeal and the troops came from so many different units, however, that it was difficult to tell just who or what was committed. Many units never reached Aloui and merely became part of the withdrawal problem. Even with all the detachments, attachments, additions, and deletions, only one-third of the cavalry squadrons and two-thirds of the tank squadrons available to I Corps were used in Laos. Numerically, this employment amounted to five tank squadrons and six armored cavalry squadrons.

Faced with superior enemy forces, the I Corps commander decided to withdraw. Although units attempted to evacuate the landing zones in an orderely fashion, constant enemy pressure caused several of the sites to be abandoned and forced the defenders to make their way overland to more secure pickup zones. Several units had considerable difficulty breaking away from the pursuing enemy and were lifted out only after intense tactical air, artillery, and aerial rocket preparation.[15] By 21 March the 1st Infantry Division had completely withdrawn from Laos and major elements of the airborne division had been lifted out.

The I Corps commander ordered the 1st Armor Brigade to withdraw on 19 March. He further allocated two U.S. air cavalry troops to the airborne division to cover the move. With the 11th Armored Cavalry Regiment as rear guard the 1st Armor Brigade began its withdrawal on time, but the brigade received no air cavalry support. Both troops had been diverted by the airborne division to support airborne battalions elsewhere.

At a stream crossing halfway between Aloui and Landing Zone Alpha, the armored column was ambushed by a large North Vietnamese force. The unit in front of the 11th Armored Cavalry abandoned four M41 tanks in the middle of the stream, where they

15 Realizing the safety factors involved in the use of supporting fire, the enemy employed "hugging" tactics to avoid the fire. They often stayed as close as 20—30 meters to friendly units. The only aerial support accurate enough that close was the helicopter gunship.

completely blocked the withdrawal route. The airborne infantrymen refused to stay with the cavalry and continued east down the road. The armor brigade commander was informed of the situation but sent no reinforcements or recovery vehicles to clear the crossing. Troopers of the 11th continued to fight alone, and after three hours succeeded in moving two of the abandoned tanks out of the way. The cavalry then crossed, leaving seventeen disabled vehicles to the west of the stream. The North Vietnamese immediately manned the abandoned vehicles, which they used as machine gun positions until tactical air strikes destroyed them on 25 March. What had begun as an orderly withdrawal was rapidly becoming a rout.

The armor brigade reached Landing Zone Alpha on 20 March, regrouped, and pushed on, still without benefit of air cavalry. The next morning the brigade, with the 11th Armored Cavalry leading, was again ambushed, this time three kilometers east of Fire Support Base Bravo. In the midst of the firefight, an air strike accidentally hit the Vietnamese column with napalm, killing twelve and wounding seventy-five. The brigade withdrew west to regroup.

By that time the armor brigade had lost approximately 60 percent of its vehicles, and when a prisoner reported that two North Vietnamese regiments were waiting farther east along Route 9 to destroy it the armored force turned south off the road. The airborne division, also aware of the prisoner's statement, had meanwhile airlifted troops north of Route 9 and cleared the ambush site. The armor brigade, unaware of the airborne action, found a marginal crossing over the Pon River, two kilometers south of Route 9. The brigade recrossed the river twelve kilometers to the east and reached Vietnam through the positions of the 1st Battalion, 77th Armor.

The withdrawal of the 1st Armor Brigade is perhaps the most graphic example of the poor coordination between major commands throughout LAM SON 719. When the brigade left Route 9, less than 5 kilometers from Vietnam via road, it was forced to make two river crossings because its commander was not told that the road had been cleared. It was this lack of coordination at the highest levels, and the apparent lack of concern for the armored forces, that contributed to the poor performance of armor.

In Operation LAM SON 719, which officially ended on 6 April 1971, South Vietnamese armor did not appear to advantage. In a static role at Aloui, armor proved no more dynamic than a pillbox, and became a liability requiring additional forces for its security. Command and control problems at all levels were evident, and plagued the operation from the start. A small amount of armor was committed at first, and reinforcement was piecemeal. None of this, however, excused the performance of some armored units which, especially during the withdrawal, simply abandoned operational vehicles in their haste to get back to safety.

Some good did come from LAM SON 719. For example, it helped to delay major enemy operations for the remainder of 1971. The intelligence gained concerning the North Vietnamese pipeline and trail network in Laos was used for planning future bombing raids.[16] The operation allowed the South Vietnamese forces to use U.S. aviation and artillery support without the assistance of American advisers, and thus paved the way for the South Vietnamese Army's complete operational control of U.S. aviation and artillery in mid-summer of 1971.

Before this operation, the South Vietnamese infantry had little or no antitank training, but the presence of enemy armor during LAM SON 719 led to greater emphasis on antiarmor techniques and instruction in the use of the M72 light antitank weapon. Both sides in LAM SON 719 lost heavily in men and equipment and there was no clearcut victory, but psychologically the Vietnamese armored forces had received a hard blow.

Cuu Long 44–02

One other South Vietnamese armored operation in 1971 was significant, although it was not widely publicized. For one reason, since it occurred at almost the same time as LAM SON 719, it was lost in the glare of reporting that operation. For another, penetration into Cambodia, the deepest of the war, made it politically sensitive. The operation was staged because the North Vietnamese had cut Route 4, the only supply road in Cambodia between Phnom Penh, the capital, and the port of Kampong Som: the Cambodian government had requested South Vietnamese assistance in reopening it.

Operation CUU LONG 44–02 began on 13 January 1971, as the 4th Armor Brigade with the 12th and 16th Armored Cavalry Regiments, three Ranger battalions, an artillery battalion, and an engineer group, moved 300 kilometers from Can Tho to Ha Tien in fourteen hours. For the next two days, the brigade pushed north along Routes 3 and 4. The first enemy encountered had set up an

---

16The 2d Squadron, 17th Cavalry (Air) with the Hac Bao Company conducted a series of deep, successful raids in Cambodia and Vietnam over a five month period using this information. The techniques developed in these raids are worthy of study for future application.

ambush that the 16th Armored Cavalry Regiment literally blew away by charging on line.

A second ambush farther north against the 12th Armored Cavalry also failed. The enemy tried to isolate the lead squadron by destroying the first and last vehicles. The lead commander, however, kept his flaming vehicle moving and his machine gun firing. Hit three times and burning, the armored cavalry vehicle continued north for about 150 meters before it blew up, killing the crew. This heroic effort prevented the column from being trapped on the road and allowed the cavalry to get out of the enemy firing lanes. The Ranger battalion behind the cavalry squadron stopped and opened fire. The ambushers were now in a deadly cross-fire between the cavalry and the Rangers. Two U.S. aerial fire teams sealed off the enemy escape routes.[17] When the smoke cleared, 200 of the enemy lay dead, and seventy-five weapons, including two 75-mm. recoilless rifles and three heavy machine guns, had been captured. The 12th Armored Cavalry Regiment lost five killed, twenty wounded, and three tracks destroyed.

On 17 January Cambodian forces, with Vietnamese Marine Corps support, fought to the outskirts of the Pich Nil Pass and secured it, while the armor brigade secured Route 4 as far north as Route 18. After helping the Cambodians set up strongpoints, the 4th Armor Brigade withdrew toward South Vietnam, arriving by 25 January.

On several occasions, Vietnamese armored units had conducted bold operations deep into enemy territory, in both Laos and Cambodia. Three major operations, CUU LONG 44–02, LAM SON 719, and TOAN THANG 01–71, all took place simultaneously, with the South Vietnamese hoping to keep the initiative gained in 1970. Of the three operations, only CUU LONG 44–02 can be regarded as a success, and as a result Military Region 4 remained one of the most secure areas in South Vietnam. The other two operations demonstrated that a parity existed in South Vietnamese-North Vietnamese strength. The year 1971 was not successful for either side; it ended with the Viet Cong and North Vietnamese as strong, if not stronger, than the South Vietnamese.

---

[17] U.S. aviation units in Military Region 4 provided gunship support from bases in the region and from the deck of the USS *Cleveland* which was cruising in the bay near Kampong Som.

CHAPTER VIII

The Enemy Spring Offensive of 1972

The largest military operation of 1970–1971 in Vietnam was not a combat operation but the redeployment of over 300,000 American troops. Although the withdrawal was governed to some extent by the fluctuating intensity of combat operations, it moved as relentlessly as an avalanche. Of the 543,000 American troops in Vietnam in April 1969, 60,000 had left by the end of that year. In accordance with the planning guidance of General Abrams, only a few small armored units were among those withdrawn.

In 1970, when 139,000 American troops—including the first armored battalions redeployed—returned to the United States, the concentration of U.S. armored battalions climbed to forty-six percent of the combat units that remained in Vietnam. The pace of the withdrawal slowed considerably during the Cambodian expeditions but after August 1970, when U.S. operations in Cambodia ended, large numbers of units were withdrawn. The highest percentage of armored units leaving Vietnam came from Military Region 2, where enemy activity was almost negligible. By the end of 1970, the 1st Squadron, 10th Cavalry, was the only armored unit left in the region. The primary mission of this unit was still road security.

As the number of American troops in Vietnam decreased, the Army of the Republic of Vietnam became the dominant force and began to assume operational control of U.S. units. One such American unit was the 2d Squadron, 11th Armored Cavalry Regiment. Late in April 1971 increasing enemy resistance to Rome plow operations along Route 1, northwest of Saigon, made it plain that mechanized forces would be needed if the project was to proceed on schedule. Accordingly, the security of the entire Rome plow operation in Hau Nghia Province was placed under the control of the 2d Squadron, 11th Armored Cavalry, which reported directly to the Vietnamese province headquarters.

Throughout 1971 however, withdrawal was the U.S. mission, and 177,000, or 53 percent, of the Americans departed. By year's end, troop strength was 158,000, the lowest since 1965. Fifty-four percent, of the remaining U.S. combat battalions were armored units—four air cavalry and two armored cavalry squadrons as well as many separate ground and air cavalry troops. Armored units slated for redeployment were withdrawn from various parts of the country during 1971, leaving armored strength evenly distributed from north to south.

A lull in the Vietnam conflict during the first months of 1972, as U.S. forces continued to disengage, was encouraging. Of all American units remaining in Vietnam, only air and ground cavalry, performing reconnaissance and security missions, continued to encounter the enemy with any degree of frequency. The immediate effect of the withdrawal schedule was that most units were relegated to administrative status, or, at best, to a local security role. While definite restrictions limited the nature and scope of American participation in ground combat, air cavalry units were under less restraint. U.S. air cavalry continued to perform its reconnaissance and security missions with little regard for the departure of other U.S. organizations.

All American armored and air cavalry squadrons still in Vietnam at the end of 1971 were ordered to redeploy before 30 April 1972. The last U.S. ground cavalry unit to conduct operations was Troop F, 17th Cavalry, which left Da Nang on 6 April. The 1st Squadron, 1st Cavalry, which began its return on 10 April, was the last ground cavalry unit to leave Vietnam. Troop D, the 1st Squadron's organic air cavalry troop, was redesignated Troop D, 17th Cavalry, and assigned to Da Nang. It remained in Vietnam until after the cease fire and supported the Army of the Republic of Vietnam ground units in Military Region 1.

As the air cavalry squadrons departed, they left behind separate air cavalry troops as the last vestige of U.S. combat strength. Thus, Troop C, 16th Cavalry, was the sole air cavalry unit in Military Region 4. When the 7th Squadron, 17th Cavalry, left on 29 April—the last air cavalry squadron to redeploy from Vietnam— it left behind two separate air cavalry troops, Troop H, 10th Cavalry, and Troop H, 17th Cavalry. These two troops operated in Military Region 2 until 25 February 1973 and were the last U.S. combat units to leave Vietnam. Air cavalry units with the primary mission of supporting South Vietnamese Army forces were the only active Army combat units in Vietnam in 1972.

Point and Counterpoint

The Vietnamese lunar New Year which began in February 1972 was the Year of the Rat. Like the rat, the enemy assumed a low profile during the first weeks of 1972. By day enemy forces avoided direct confrontation with free world forces, and by furtive scurrying at night built up extensive hoards of supplies and equipment in border staging areas.

As early as November 1971, the intelligence community, the government of South Vietnam, and U.S. and South Vietnamese Army commanders anticipated a significant enemy offensive in 1972, expecting the main effort to be made in mid-February. The military objectives of the offensive were not known, but intelligence sources reported that its goal would be the destruction of the Army of the Republic of Vietnam. Not even at the highest levels of government, however, was a major shift in enemy tactics expected.

That the North Vietnamese were capable of a large-scale offensive equal to the *Tet* attack of 1968 was apparent. The possibility that enemy armor would be a threat was considered insignificant, however, since only three times during the Vietnam War had the North Vietnamese Army employed armored vehicles in combat. Tanks in the North Vietnam inventory were estimated at more than 300, and were thought to be organized in three regiments. Two regiments were known to be operating in Laos and Cambodia, where intelligence reports indicated that they would remain.[1] A report prepared by Headquarters, U.S. Army, Pacific, in February 1972, surmised that terrain, logistics, and free world firepower would limit the size and general location of armor-supported assaults. Sustained operations with tank units larger than a company were considered impossible without establishment of large fuel and supply caches in the border areas. The activity required to establish these stockpiles would reveal enemy intentions and subject the forces and supplies to devastating air attacks.

In early March enemy offensive action in Military Regions 1 and 2 increased and involved larger units. Continuing intelligence from sensors and other sources in the northern part of Military Region 1 focused attention on the A Shau valley, where free world forces had not operated since 1970. (*Map 16*) Road construction, enemy logistical troops, heavy artillery, antiaircraft weapons, and tanks were detected with increasing frequency. The extent and intensity of antiaircraft fire almost halted our air reconnaissance in the mountainous areas west of Fire Support Base Bastogne. The South Vietnamese 1st Infantry Division looked on this buildup as a threat to Hue, and on 5 March started an operation called

1 Tread marks discovered by scout helicopters of Troop F, 4th Cavalry, indicated that a number of tracked vehicles had moved out of a rubber planatation northwest of Krek, Cambodia, and apparently staged a practice assault against an abandoned fire support base.

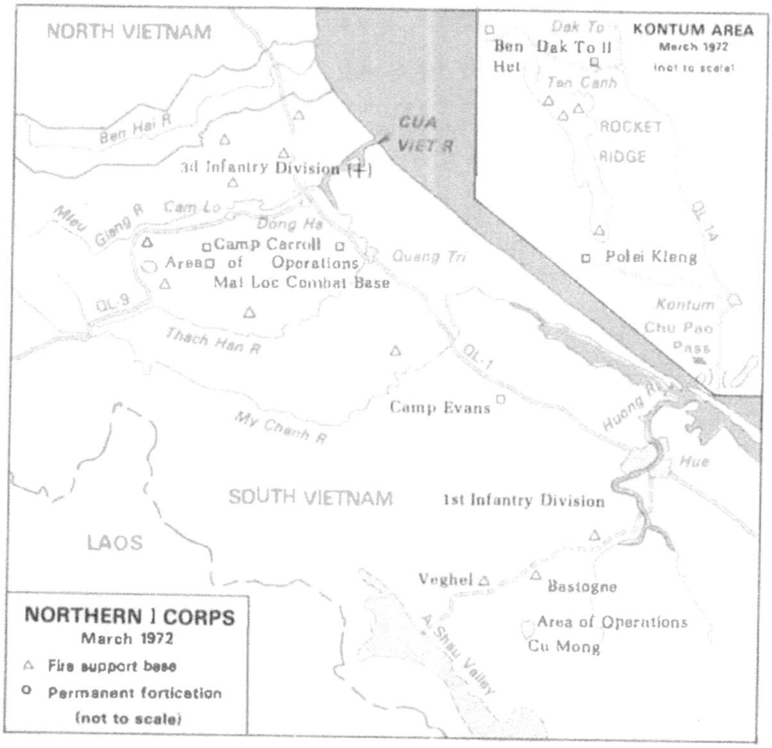

MAP 16

LAM SON 45–72 to destroy a logistical base near Cu Mong Mountain.

Deliberately launched before the end of the wet season, this major undertaking was planned as a joint airmobile and ground attack, but overwhelming antiaircraft quickly compelled the South Vietnamese Army to advance only on the ground. Heavy fighting took place early, and casualties from intense artillery fire and daily B–52 bomber strikes forced the North Vietnamese Army to commit major reinforcements from the 324B Division. A thrust by the 7th Armored Cavalry Regiment and an infantry regiment in the vicinity of Fire Support Base Veghel stirred up a hornet's nest. In the ensuing struggle the armored cavalry lost two troops. The extent and ferocity of the fighting convinced the South Vietnamese that the enemy intended to attack Hue in force. Pulling back to more favorable terrain near the Bastogne base, South Vietnamese soldiers fought continuously with North Vietnamese units well into May.

In Military Region 2 reports on enemy tank sightings in the border areas persisted but remained unconfirmed. During March the Air Force destroyed significant qualities of stockpiled enemy supplies detected by the American air cavalry. South Vietnamese cavalry units continued to perform route security missions but otherwise staged no offensive operations.

The 20th Tank Regiment

Northern Military Region 1, a critical area bordered by North Vietnam and Laos, was protected in the summer of 1971 by South Vietnamese infantry, three South Vietnamese armored cavalry squadrons, and the U.S. 1st Brigade, 5th Infantry Division (Mechanized), scheduled to leave the country in August.[2] Analysis of terrain, the probable enemy threat, and enemy armored actions during LAM SON 719 made it clear that armored units would continue to be needed in this area. Consequently, the Vietnamese Joint General Staff authorized on 31 July 1971 the formation of the 20th Tank Regiment. Equipped with M48A3 tanks, it was the first South Vietnamese tank regiment

Tailored specifically to fit the needs and capabilities of the South Vietnamese Army, the 20th Tank Regiment had an unusual organization. During LAM SON 719, armored vehicles had proven vulnerable to individual antitank weapons when not protected by infantry. The Joint General Staff had therefore directed an addition to the regiment, a 270-man armored rifle company. Ninety riflemen were assigned to each tank squadron and were to ride on the outside of the tanks, providing local security.

Other changes in the tank regiment's organization and equipment included the addition of tracked M548 ammunition and fuel cargo vehicles, elimination of the regimental scout platoon, for which a five-vehicle security section was substituted, and elimination of the armored vehicle-launched bridge section and all infrared fire control equipment. Later six xenon searchlights per squadron were authorized after advisers questioned the wisdom of limiting $15 million worth of fighting equipment to daytime use by refusing

[2] To increase Vietnamese combat power and to fill the gaps left by redeploying U.S. units, portions of the South Vietnamese 1st and 2d Infantry Divisions and local units of the Regional and Popular Forces were combined to create the 3d Infantry Division. The 11th Armored Cavalry, drawn from the 1st Armor Brigade, became the division's organic cavalry regiment.

to spend $300,000 on searchlights. Unfortunately, the decision to dispense with the vehicle-launched bridge section was not reconsidered, and lack of bridging during the enemy offensive proved a major factor in the loss of tanks.

Training for the 20th Regiment began at Ai Tu near Quang Tri City, but proceeded slowly because of many problems, particularly in maintenance. About 60 percent of the tanks received by the regiment had serious deficiencies beyond the repair capability of the tank crews. Repair parts and technical manuals were missing and the language barrier prevented U.S. instructors from communicating adequately with the Vietnamese crewmen.

On 1 November a gunnery program based on U.S. tank standards got under way. Unfortunately, the inexperienced tank crews had difficulty in comprehending the integrated functioning of the rangefinder and ballistic computer. In fact the Vietnamese language could come no closer to the term ballistic computer than to translate it "adding machine." Partly because of their experience with the M41 tank, which had no rangefinder, Vietnamese commanders at first could not be convinced of the rangefinder's value. Rapid troop turnover and manpower shortages also adversely affected crew performance. Training therefore made slow headway, with many reversions to basic lessons. By 25 January gunnery training ended, with 41 of 51 available crews qualifying, using test criteria as rigorous as those used for U.S. units.

Unit tactical training began in the foothills west of Quang Tri City on 1 February and was judged successful in its later stages. A recurring problem during tactical testing was the Vietnamese inclination to disregard maintenance before, during, and after an operation. Continued emphasis on maintenance resulted in some improvement, but standards remained below acceptable levels, even after the unit completed its training.

The regiment's final tactical test, a field training exercise, was to be conducted by the South Vietnamese Armor Command along U.S. lines, with the proviso that any portion not completed correctly was to be repeated. Several problems delayed the exercise past its scheduled starting date of 13 March. Poor weather during the gunnery phase, the necessity for some tactical retraining at the troop level, and the lack of M88 recovery vehicles and M548 tracked cargo vehicles to carry fuel combined to cause setbacks. Finally, after devoting several days to vehicle maintenance, the regiment began its training test on 27 March. Within a few days the exercise was transformed into the ultimate test—survival on the field of battle.

Attack Across the Demilitarized Zone

By the end of March 1972, South Vietnamese defenses in Military Region 1 were arranged in a roughly crescent-shaped pattern of fire support bases in northern Quang Tri Province, with the majority of forces oriented to the north and west in the vicinity of the Demilitarized Zone. (*See Map 16.*) To the south and stretching westward to Highway 1 from the sea were a number of small Regional Forces outposts commanded by province and district chiefs. From Highway QL-1 west and south, roughly paralleling the mountains, was the newly formed South Vietnamese 3d Infantry Division, bolstered in the west by the Vietnamese Marine division. Deployed with the regiments of the 3d Infantry Division was its organic cavalry, the 11th Armored Cavalry. Farther to the south, guarding exits from the A Shau valley and the approaches to Hue, was the South Vietnamese 1st Infantry Division with its organic 7th Armored Cavalry Regiment.

In the early morning hours of 30 March devasting rocket, mortar, and artillery fire fell on every fire support base in Quang Tri Province. The bombardment continued all day, and late in the day the northernmost bases reported North Vietnamese Army tanks and infantry moving south across the Demilitarized Zone. Major General Frederick J. Kroesen, Jr., Deputy Commander, U.S. XXIV Corps, described the action:

> The artillery offensive was followed by infantry and armor attacks in the east across the Ben Hai River following the axis of QL (Route) 1 in the west toward the district capital of Cam Lo and Camp Carroll. Elements of the 304th and 308th Divisions, three separate infantry regiments of the B5 Front, two tank regiments, and at least one sapper battalion were later identified among the attacking forces. Initially then, the enemy concentrated a numerical advantage of more than three to one over the defending 3d Division and attacked forces which were disposed to counter the infiltration and raid tactics heretofore employed by the NVA in the DMZ area.

The tactical situation on 30 March was confused; the 3d Infantry Division received vague and conflicting reports from fire bases at an astonishing rate. Most disconcerting were accounts of the ferocity and widespread nature of the attacks. Just before noon the 20th Tank Regiment received a frantic message from Headquarters, Military Region 1, ordering it to return to Quang Tri City. Since no explanation was given, Major General Nguyen Van Toan, Chief of Armor, and his American adviser, Colonel Raymond R. Battreall, Jr., flew to Quang Tri City to see General Vu Van Giai, the South Vietnamese 3d Division commander. There they learned that the western fire bases near the Demilitarized Zone had been overrun in a prelude to what was apparently a major enemy offensive. Since the main attack had not yet been identified, and since no one was sure where the tank regiment would be of the most value, General Toan persuaded General Giai not to commit the 20th Regiment prematurely but to hold it as a division reserve or for use as a counterattack force. He also convinced General Giai that he should permit the unit to stand down for maintenance before its commitment. With that determined, the regiment, then conducting its final coordinated assault phase of the training exercise, completed the assault, did a right flank on the objective (out of line formation and into a column), and, without stopping, returned to Ai Tu Combat Base.

The Rock of Dong Ha

By early morning of 1 April most of the outlying fire bases along the Demilitarized Zone and in western Quang Tri Province had been evacuated or overrun, leaving no friendly positions north of the Mieu Giang and Cua Viet rivers. Poor weather prevented air support and contributed to the relative ease with which the enemy pushed back the South Vietnamese. The North Vietnamese forces advanced south with impunity. By late afternoon on 1 April Mai Loc and Camp Carroll, south of the Mieu Giang River, were under heavy attack.

Frantically redeploying the three infantry regiments, one cavalry regiment, and two Vietnamese Marine brigades at his disposal, General Giai established a defensive line along the south bank of the Mieu Giang. In an effort to stabilize the situation, he committed the 20th Tank Regiment on the morning of 1 April with the mission of relieving the embattled 11th Armored Cavalry Regiment and attached infantry units then fighting around Cam Lo, along National Highway 9. (*Map 17*) After joining a South Vietnamese Marine battalion, the tank regiment moved north from Ai Tu along Highway 1 toward Dong Ha.

Poor traffic control and refugees congesting the route forced the tank regiment to move cross-country to the southwest of Dong Ha, and in so doing it surprised and routed an enemy ambush along Highway QL-9. Prisoners taken during this action were dismounted members of a North Vietnamese tank unit whose mission was to seize and man South Vietnamese armored vehicles expected to be captured in the offensive. With its forty-four operational tanks, the 20th Tank Regiment moved on toward Cam Lo, which was burning. As darkness approached, the unit set up a defensive

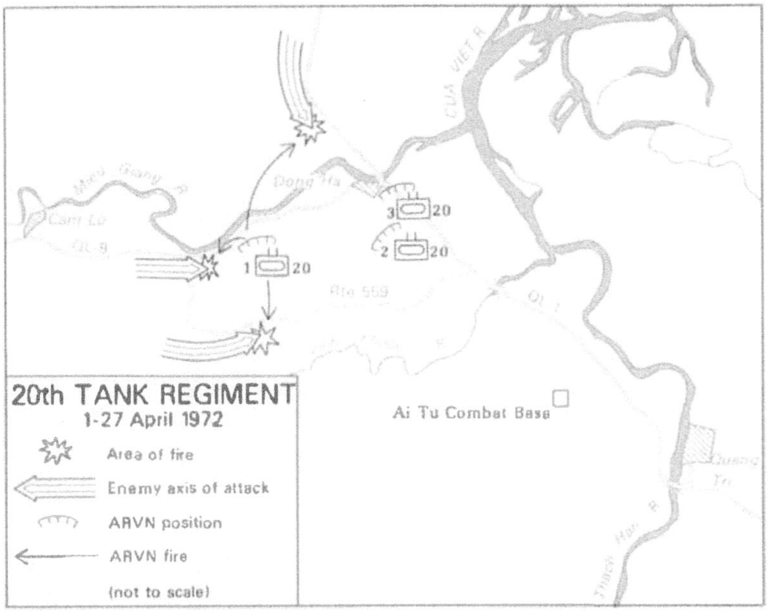

MAP 17

position southeast of Cam Lo Village, withstanding enemy probes throughout the night.

At daybreak on Easter Sunday, 2 April, the 20th Tank Regiment received reports that a large North Vietnamese tank column was moving south across the Ben Hai River toward the bridge at Dong Ha. About 0900 the commander, Colonel Nguyen Huu Ly, received permission to move to Dong Ha, then north across the bridge to engage the enemy forces. When he reached the town he found enemy infantry already occupying positions on the north bank of the Mieu Giang River that prevented his crossing the bridge. He deployed the regiment around the town of Dong Ha, with the 1st Squadron in a blocking position on the high ground about three kilometers to the west, the 2d Squadron to the south, and the 3d Squadron defending positions within the town to prevent enemy elements from crossing the bridge.

About noon men of the 1st Squadron, from their vantage point on the high ground to the west, suddenly observed a North Vietnamese tank and infantry column moving south along Highway I toward Dong Ha. Moving their tanks into concealed positions, they waited as the enemy tanks moved closer. At a range of 2,500 to

3,000 meters, the South Vietnamese tankers opened fire, quickly destroying nine PT76 tanks and two T54 tanks. The North Vietnamese unit, which by its column formation showed that it was not expecting an attack, was thrown into confusion. Unable to see their adversaries, the North Vietnamese crewmen maneuvered their tanks wildly as the South Vietnamese tank gunners destroyed them one by one. The accompanying infantry dispersed, and the surviving T54 tanks turned and headed north without firing a single shot. The South Vietnamese regimental headquarters, monitoring the North Vietnamese radio net at that time, heard the enemy commander express surprised disbelief at losing his tanks to cannon he could not see.

The steady deterioration of the tactical situation around Dong Ha was arrested by the arrival of the 1st Armor Brigade headquarters. Although the brigade headquarters had been in the area solely to monitor the 20th Tank Regiment's training exercise, it was a well trained organization, possessing the armored vehicles and radios needed by General Giai to establish control of the scattered forces and direct the defense he hoped to establish at Dong Ha. General Toan had urged its employment, and on the afternoon of 2 April the brigade, under 3d Division control, assumed command of all armored, infantry, and Marine forces in the Dong Ha area. Its units included the 20th Tank Regiment, two squadrons of the 17th Armored Cavalry Regiment, the 2d and 57th Regiments of the 3d Infantry Division, the 3d Battalion of the 258th Marine Corps Brigade, and the survivors of the 56th Regiment from Camp Carroll.[3]

The bridge spanning the Mien Giang River at Dong Ha afforded the enemy the opportunity to cross the river unimpeded and then drive straight south to Quang Tri City. Before the armor brigade headquarters arrived, the 3d Division engineers had made two unsuccessful attempts to destroy the bridge with explosive charges. When Colonel Nguyen Trong Luat, the 1st Armor Brigade commander, arrived he decided to leave the bridge intact for the time being, since the enemy had been stopped and the armor brigade forces were holding. Colonel Luat was preparing to make a counterattack to the north across the bridge when the bridge charges detonated and dropped the near span, putting an end to any counterattack plans.

[3] At 0800 on 2 April, after successfully defending the camp against three major enemy assaults, the 56th Regiment inexplicably gave way. The loss of Camp Carroll made the position at Mai Loc untenable, and the Marine units at that location were withdrawn over the next three days.

Other enemy forces continued to move south toward Dong Ha on the afternoon of 2 April, engaged first by limited tactical air strikes and then by artillery, mortar, and tank fire. A large search and rescue effort had been launched for the crew of a U.S. aircraft downed near Cam Lo. The U.S. Air Force temporary no-fire zone was twenty-seven kilometers in diameter, encompassing nearly the entire combat area and South Vietnamese Army defenders were unable for several hours to call for artillery support or tactical air strikes against the onrushing North Vietnamese Army. The enemy therefore had an opportunity to advance artillery, tanks, and infantry until 2200, when the restriction was lifted.

During the next several days, enemy activity was relatively light, with sporadic attacks by fire and numerous small ground actions. The North Vietnamese artillery fire was extremely accurate, and although South Vietnamese units moved frequently to avoid the shelling the enemy seemed to be able to locate new positions very quickly. On 3 April a North Vietnamese artillery observer in a South Vietnamese officer's uniform and driving a South Vietnamese vehicle with radios was captured south of Dong Ha. The observer had papers supporting several identities, and spent his time driving throughout the area spotting and adjusting artillery fire for North Vietnamese guns near the Demilitarized Zone. Although South Vietnamese units conducted attacks to eliminate pockets of resistance south of the Mieu Giang River, the pressure from the north remained intense.

The next tank combat occurred on the 9th when all three squadrons of the 20th Tank Regiment fought enemy armor. The 1st Squadron, shifted several kilometers west of Dong Ha six days earlier, occupied high ground overlooking an important road junction along National Highway 9. Again the tank gunnery training paid dividends as the tankers engaged an infantry unit supported by ten tanks at ranges up to 2,800 meters. A few answering shots fell short, and the enemy tanks scattered, several bogging down in the rice paddies near the road. Eventually eight were destroyed. In all, the regiment destroyed sixteen T54 tanks and captured one T59 that day, in turn suffering nothing more than superficial damage to several M48's.

For the next two weeks the South Vietnamese carried out clearing operations interrupted by frequent engagements with North Vietnamese armor and infantry which normally withdrew in the late afternoon. Nights were punctuated by artillery, mortar, and rocket attacks on South Vietnamese positions throughout the area. The defensive lines established on 2 April continued to hold,

NORTH VIETNAMESE T59 TANK *captured by South Vietnamese 20th Tank Regiment south of Dong Ha.*

and on 11 April the 1st Armor Brigade was augmented by the arrival of the 18th Armored Cavalry Regiment from Military Region 3. By 14 April the 3d Division controlled five regimental-size South Vietnamese task forces, including units of the 4th, 11th, 17th, and 18th Cavalry Regiments and the 20th Tank Regiment.

On 23 April, several kilometers west of Dong Ha, the 2d Squadron of the 20th Tank Regiment was attacked by an infantry-tank force using a new weapon. For the first time the enemy employed the Soviet AT3 Sagger wire-guided missile, destroying an M48A3 tank and an armored cavalry assault vehicle. A second assault vehicle was damaged. At first the South Vietnamese Army tankers seemed fascinated by the missile's slow and erratic flight. Through trial and error, however, the troops soon learned to engage the launch site of the AT3 with tank main gun fire and to move their vehicles in evasive maneuvers.

Heralded by massive artillery attacks with 122-mm. rockets and 130-mm. guns, on 27 April a new enemy offensive began against South Vietnamese Army positions all along the Mieu Giang–Cua Viet River defense line. The barrage was quickly followed by violent attacks by enemy infantry and armor, met by equally determined resistance on the part of the South Vietnamese defenders. The 3d Squadron, 20th Tank Regiment, supporting the 5th Ranger Group, received the brunt of the attack and was soon heavily engaged. By midmorning all officers of the 3d Squadron had been killed or wounded, and three M48A3 tanks had been destroyed by Sagger missiles.

All along the defensive line, units were being overrun or pushed back. Forced to yield ground, Ranger and tank elements gradually withdrew to the southeast. Although losses were heavy on both sides, the numerically superior North Vietnamese continued their drive, and by nightfall had pushed almost four kilometers south of Dong Ha. In the early morning of 28 April, the 20th Tank Regiment had eighteen operational M48A3 tanks. During the South Vietnamese withdrawal the accurate gunnery of the 3d Squadron cost the North Vietnamese five T54 tanks.

At that point the South Vietnamese found large enemy forces to their rear and for the armored units the withdrawal became an attack to the south. The 2d Squadron of the tank regiment, attacking south to secure the bridge over the Vinh Phuoc River at midmorning on the 28th, was badly battered in an enemy ambush. The commander lost control of his unit and the surviving vehicles, after crossing the bridge, continued to the south in disarray.

It was then obvious to Colonel Luat that 1st Armored Brigade units were threatened with encirclement, so the entire force began moving south. All along the way fighting was heavy for the next two days. The terrain as well as the enemy took its toll of vehicles. At the Vinh Phuoc River seven vehicles were stranded on the north shore when the bridge, struck by enemy artillery fire, collapsed.[4] Farther south at the Thach Han River near Quang Tri City, the bridges were already destroyed. Two tanks were lost there in fording the river on the 30th.

By then the tank and cavalry units were beginning their fifth day of almost constant fighting. South of Quang Tri resupply of fuel and ammunition was nonexistent as the armored force continued its attack. Forced from the highway by a determined enemy, the tanks and assault vehicles moved cross-country, falling victim to the many rice paddies, canal crossings, and streams as well as the antitank rockets and artillery. On the first day of May the vehicles began to run out of gas.

[4] Two M113 command vehicles, approaching the bridge and unable to stop in time, ran up the sagging, inclined span and hurtled through the air for nearly a full vehicle length, landing safely in the part of the bridge that was still intact. Seconds later the span collapsed.

Finally, on 2 May, having fought their way through the last enemy units, the battered survivors of the armor command, intermingled with the remnants of other army units, reached Camp Evans at midafternoon. Only armored cavalry assault vehicles were left; the cavalry regiments and the tank regiment had lost all their tanks. The once proud 20th Tank Regiment was reduced to a demoralized, dismounted, and defeated unit. Employed primarily in a static, defensive role in frontline areas, the unit had steadily lost men and equipment without receiving replacements. Although vastly outnumbered, cavalry, infantry, tank, and Marine units of the 1st Armor Brigade, as well as tenacious Regional Forces and Popular Forces to the east, had succeeded in slowing the momentum of the massive North Vietnamese invasion. With assistance from U.S. and Vietnamese tactical air forces, they provided the resistance that delayed the enemy until enough reinforcements could be brought up to halt the offensive.

The Enemy Attack in Military Region 2

In Military Region 2 essentially the same preliminary enemy activity took place that had been seen in Region 1. As early as mid-December 1971, air cavalry reconnaissance confirmed reports of large-scale troop movements in the border areas of Cambodia and Laos. The 7th Squadron, 17th U.S. Cavalry (Air), engaged two tanks and sighted four others west of Kontum on 25 January 1972, but the advisory staff of Military Region 2 failed to recognize the importance of the reports. Enemy units were reportedly staging for a campaign against population centers like Kontum and Pleiku, to be supported with attacks by local Viet Cong units in the coastal lowlands designed to draw off South Vietnamese forces.

To counter what appeared to be a growing threat to Kontum, the regional commander, Lieutenant General Ngo Dzu, moved the bulk of the 22d Infantry Division to Tan Canh and Dak To II, including the division's organic 14th Armored Cavalry Regiment as well as reinforcing elements from the 1st Squadron (tank), 19th Armored Cavalry Regiment. (*See Map 16, inset.*) Most of the M41 tanks of both regiments were in position at Ben Het by order of the 22d Division commander, against the advice of the 2d Armor Brigade commander, who argued that cavalry should be used in a mobile role. In mid-March encounters with large enemy units along Rocket Ridge and near Ben Het, intelligence and prisoner reports indicated that a full-scale offensive supported by tanks was being planned for the period April to September.

By 23 April Tan Canh and Dak To II were effectively sur-

DISABLED M48 OF THE SOUTH VIETNAMESE 20TH TANK REGIMENT. *Road wheels received hit from rocket near My Chanh River.*

rounded by North Vietnamese forces, and in the afternoon of that day an M41 tank of the 1st Squadron, 14th Cavalry, returning through the main gate of the compound at Tan Canh, was hit and destroyed by a Sagger missile. Within moments additional missiles hit the reinforced bunker that served as the tactical operations center for the 22d Infantry Division, killing or wounding several members of the staff. Shortly thereafter the remaining M41 tanks within the Tan Canh compound were also destroyed by missiles.

At 2100 a column of eighteen North Vietnamese tanks was reported moving south toward Dak To. Shortly after dawn on 24 April, enemy tanks and infantry attacked two sides of the Tan Canh perimeter, causing confusion and panic among South Vietnamese Army support troops. These troops fled, followed not long after by the remaining South Vietnamese Army combat troops, who had been surprised and demoralized by the sudden onslaught of the T54 tanks.

At Dak To II, five kilometers to the west, North Vietnamese infantry and T54 tanks penetrated the perimeter in several places. By midafternoon both bases were in enemy hands, and the scat-

UH–1B HELICOPTER WITH TOW MISSILES *lifts off strip at Pleiku.*

tered, disorganized remnants of the 22d Infantry Division were evading capture in the jungle.

The North Vietnamese spent the next several days consolidating their positions and tallying up the massive arsenal South Vietnamese units had abandoned intact: twenty-three 105-mm. howitzers, seven 155-mm. howitzers, a number of M-41 tanks, and about 15,000 rounds of artillery ammunition.

On 25 April South Vietnamese forces abandoned their precariously positioned fire bases along Rocket Ridge, thereby allowing the enemy to proceed unhindered down Highway QL-14 toward Kontum City. The city, meanwhile, was being hastily fortified by elements of the South Vietnamese 23d Infantry Division, which had been given control over all forces in Kontum Province. Between 25 April and 9 May, enemy attacks were made against Ben Het and Polei Kleng Ranger camps located astride North Vietnamese supply routes. While Polei Kleng fell on 9 May, Ben Het stood firm against repeated tank and infantry attacks throughout the campaign. By this time attacking enemy units had been identified as elements of the 2d and 320th North Vietnamese Divisions, supported by the 203d Tank Regiment.

The 23d Infantry Division commander, Colonel Ly Tong Ba, who had commanded the original M113 mechanized rifle company in 1963 as a captain, now worked feverishly to prepare his troops both physically and psychologically to withstand an enemy attack spearheaded by armor. He instituted an intensive program to train division soldiers in the use of the M72, an American light antitank weapon, and tried to instill confidence in the individual soldier that he could destroy an enemy tank with the weapon. Artillery concentrations were planned and coordinated with B–52 bomber strikes, and practice counterattacks were made in areas of possible penetration. All in all, the time the South Vietnamese gained by the enemy failure to follow up the overwhelming victory at Tan Canh was put to good use.

At this time a new weapon, destined to change the complexion of armored warfare, appeared on the battlefield in Military Region 2. On 24 April the 1st Combat Aerial TOW Team arrived in Saigon. Organized to participate in an experiment at Fort Ord, California, the team was alerted on 15 April for deployment to Vietnam to conduct an extension of the original test under combat conditions. The team consisted of three crews, two UH–1B helicopters mounting the XM26, a tube-launched, optically tracked, wire-guided missible (TOW), and technical representatives from Bell Helicopter, Hughes Aircraft, and the U.S. Army Missile Command. When they arrived in Vietnam, the team moved to Pleiku for gunnery training, and after 2 May made daily flights in search of enemy armor. On 9 May, during a North Vietnamese attack on the Ben Het Ranger camp, the TOW team destroyed its first three PT76 tanks and broke up the enemy assault.

On 13 May intercepted enemy radio traffic confirmed earlier reports by air cavalry scouts that there was a large buildup of North Vietnamese armor and infantry near Vo Dinh, a signal that an all-out attack on Kontum was imminent. At dawn on 14 May the attack began. Remembering the ease with which they had taken Tan Canh, the North Vietnamese dispensed with an artillery preparation and attacked south along Highway QL-14 with two regimental-size task forces supported by tanks. A third regiment advanced from the northeast, and a fourth probed the city's southern defenses.

The 23d Infantry Division's intensive antitank training paid off on the morning of 14 May. Tank killer teams equipped with the LAW quickly destroyed most of the attacking tanks, leaving U.S. tactical aircraft and the TOW team the job of picking off the remainder, which were attempting to flee back up Highway 14. The TOW team destroyed two T54 tanks. By 0900 the attack had been repulsed, although sporadic ground probes and artillery and mortar fire continued throughout the day and night.

For the next few days, Kontum City was pounded by indirect fire and occasionally assaulted by small units. On 21 May the 23d

ACAV OF SOUTH VIETNAMESE 17TH ARMORED CAVALRY *takes up position near My Chanh for counterattack, July 1972.*

Infantry Division directed a task force composed of two Ranger groups and the 3d Armored Cavalry Regiment to clear Highway QL-14 north from Pleiku to Kontum. Clearing was successful as far as the Chu Pao Pass, about 15 kilometers south of Kontum, where it was stalled by determined enemy resistance and the loss of several vehicles.

At 0100 on 26 May the North Vietnamese Army started a second determined attempt to overrun Kontum City. Tank and infantry teams attacked from the north under cover of heavy artillery bombardment and soon made several penetrations between defending South Vietnamese elements. TOW aircraft were brought from Pleiku at first light and arrived over Kontum to find enemy tanks moving almost at will through the northern sections of the city. Tactical air and gunships could not be used without risk to friendly forces but the TOW proved ideal for the delicate task of picking off enemy tanks.

The first TOW crew, Chief Warrant Officers Edmond C. Smith and Danny B. Rowe, flew three sorties, destroying five T54 tanks and one PT76 tank, and damaging a PT76 tank. On the second sortie the TOW missile's pinpoint accuracy was demonstrated when it destroyed an enemy machine gun emplacement atop a concrete water tower. The second crew, consisting of Chief Warrant Officers Douglas R. Hixson and Lester F. Whiteis, in two sorties destroyed three more PT76 tanks, a truck, and another enemy machine gun that had replaced the gun on the same water tower.

After bombarding Kontum with artillery during the night, the North Vietnamese attacked again on the morning of 27 May with tanks and infantry. The defenders were aided by TOW-equipped helicopters, air cavalry teams, U.S. Air Force and South Vietnamese Air Force tactical aircraft, and planned B–52 strikes that pounded the attacking enemy. On the ground, South Vietnamese infantry supported by tanks from the 8th Armored Cavalry Regiment fought to dislodge the enemy from the northern part of the city, and fierce house-to-house fighting went on in the southern part. The battle continued through 30 May with neither side making visible progress. By midday on 31 May enemy troops, needing resupply and battered by several days of attack from both air and ground forces, withdrew, leaving behind nearly 4,000 dead.

The stranglehold on the supply line to Kontum was finally broken on 19 June when the 3d Armored Cavalry Regiment succeeded in reaching the city. After several weeks of bloody fighting along Highway QL-14 at the Chu Pao Pass, the 3d Cavalry turned the entrenched North Vietnamese position by maneuvering west. There could be no doubt that the enemy had suffered a resounding defeat in Military Region 2.

The Aftermath

In Military Region 3 the North Vietnamese attacks were just as fierce as those in the other two regions, but they were less successful because of massive fire support and a very determined defense by South Vietnamese infantrymen. Armored units participated in the counterattacks and relief columns but were not employed effectively as hard-hitting offensive forces. Only Military Region 4 escaped a formal attack. June, July, and August 1972 were months of reassessment and realignment of forces and ideas in South Vietnam. Scattered units were re-formed, new units were organized, and plans were made for a counteroffensive to restore national boundaries. The South Vietnamese made these preparations knowing only minimal U.S. support would be available. In keeping with previously announced redeployment schedules, American troops continued to withdraw, with the exception of Army and Air Force aviation units. The separate air cavalry troops remaining in the country participated in varying degrees during the counteroffensive period. Typical missions included bomb damage assessment, visual reconnaissance in support of South Vietnamese ground units, and routine security missions.

Initial resistance to the South Vietnamese counteroffensive was moderate in Military Region 1 but increased sharply as the enemy fell back toward Quang Tri City. Attacking forces were the Marine and Airborne Divisions, supported by armored units from the 1st Armor Brigade (the reconstituted 20th Tank Regiment and the 11th, 17th, and 18th Armored Cavalry), and some Ranger units. During the attack and capture of Quang Tri City on 16 September and afterward, small unit combined arms operations were used successfully by several commanders. Often these operations were not directed by higher headquarters but were planned and carried out by commanders at battalion level. For this reason, their successes were more noteworthy.

During the final months of 1972 both sides prepared for peace. These preparations included a massive infusion of arms and equipment by both sides. In an eleventh-hour American effort in late November, 59 tanks, 100 armored personnel carriers, and over 500 aircraft were shipped to South Vietnam.[5] North Vietnam attempted a similar buildup for North Vietnamese forces in the south, but had to contend with massive B–52 bomber raids along their supply routes.

A peace accord was scheduled to become effective in Vietnam at 0800 on 28 January 1973, after nearly five years of on-again, off-again negotiations between Hanoi and the United States. After several false starts and many months of false hopes, peace seemed to be at hand. In fact, the bloodletting was far from over.

With the 28 January cease fire, the last U.S. advisers left the country, and for the first time South Vietnamese divisions and corps were truly on their own. Not only were there no advisers to summon and coordinate the once vast U.S. tactical and strategic air power, naval gunfire, and on-call resupply but these resources themselves no longer existed. But although some South Vietnamese commanders were forced to readjust their battlefield techniques, most armored commanders had not become overly dependent on air support. The organic firepower available to armored unit commanders generally had made them more self-sufficient and self-confident than commanders of other ground units. Consequently, the departure of advisers from tank and cavalry units, which in most cases had already occurred by mid-1971, did not have much impact.

With the cease fire came an overdue change in the role of

[5] The only hindrance was demonstrators in Japan who lay down in the streets to prevent the tanks from moving from the depot to the port. Eventually an operation, much like a combat one, was mounted to move the tanks in the dead of night. It included an open telephone line to the Pentagon and was greeted with cheers in Washington, D.C., when it succeeded.

armored forces in Vietnam. Armored units had been employed in a purely tactical role as frontline troops for maneuver and fire support. In practically every operation of size or note, cavalry was there, slugging it out alongside infantry or spearheading an offensive against an enemy sanctuary. But the conventional warfare and large-scale operations initiated by the North Vietnamese during their spring offensive of April 1972 had dictated a substantial change in this employment. Continuous frontline exposure of armored units in static defensive positions soon resulted in unacceptably high losses in men and vehicles. Fortunately, there appeared to be growing awareness among high level commanders and staff officers of the need to use armored and cavalry forces as mobile reserves.

The experiences of April 1972 made it immediately apparent that using armored units in a static role, where inherent advantages of firepower, mobility, and shock effect could not be exploited, invited piecemeal destruction. The tactical situation that existed immediately before and for some time after the cease fire was ideal for the employment of armored forces as mobile reserves.

CHAPTER IX

Reflections

It is always difficult to draw up a list of lessons to be inferred from the experiences of any war. It is even more difficult, perhaps presumptive, to extrapolate the lessons of one war, and, invoking some rule of universality, correctly claim their relevance to another war—especially to one in the future. It is obvious that we do not readily learn from our own mistakes, and that we learn even less from the mistakes of others. For example, the penchant of the French for piecemeal use of armored units and how that practice worked to their disadvantage in Vietnam had been recorded. In addition, cautions against peacemeal use of armored units had been an important part of U.S. military doctrine since World War II. These words of caution came after a long and bitter struggle between a handful of American cavalrymen who saw in armored forces something more than support for dismounted infantry and American infantrymen who clung tenaciously to the idea that armored forces were merely support for infantry. But many American combat leaders, both young and old, never heeded the caution, despite the experiences of armored units in World War II. We went on to make the same mistakes again in Vietnam, with air cavalry, ground cavalry, mechanized infantry, tank battalions, and other units. We simply had not learned our lessons.

In Vietnam the cost to U.S. forces of committing armor piecemeal was not noticeably high, but on another battlefield, against a more powerful enemy—one that could capitalize on the mistake by destroying any fragmented force—the mistake could be fatal. Was it recognition that the enemy in Vietnam was unlikely to be able to destroy the fragmented forces that persuaded senior U.S. commanders to split their armored units? Or was it a serious mistake reflecting the failure of the military to learn from the past? Armor soldiers would argue for the latter—that it was a mistake, a typical and frequently repeated mistake in any war generally viewed by senior commanders as an infantry war. It was made in Korea, it was made in Vietnam. In the case of Vietnam, advice based on a considerable body of experience was available from American officers who had served as advisers to South Vietnamese armored units early in the war. For a number of reasons this advice either was not offered to the right people or was not heeded by senior officers able to influence policy and tactics in the employment of armor. We cannot afford to make this mistake again.

The second lesson of Vietnam has to do with finding the enemy, to which was closely tied possibly the most exciting development of the Vietnam era: the fielding of air cavalry. Although the problem exists to some extent in any war, in Vietnam the need to find the enemy before he could assemble and organize his forces was critical. Especially important in a future war will be an early knowledge of where the enemy has massed those weapons that will be vital to success in the battle. The special mobility of air cavalry will provide a badly needed means of reconnaissance and surveillance.

In the later stages of the war in Vietnam, when air cavalry was confronted with sophisticated enemy air defenses, it became apparent that the reconnaissance could still be performed if the commander was willing to pay the price of knocking out enemy air defenses. If information on the enemy is necessary, then the price must be paid. We must not dispense with air cavalry on the theory that it can only survive against an enemy possessing little or no air defense. The scouting mission—reconnaissance—is still critical. Air cavalry adds a new dimension to reconnaissance, one complementary to reconnaissance by ground scouts in armored cavalry units. That armored cavalry units in Vietnam were widely used as combat maneuver forces should not be allowed to obscure the fact that they are still a part of the central core of the reconnaissance team. The air cavalry-ground combination can give a much needed advantage to the force commander who uses it wisely.

In Vietnam there was considerable use of air cavalry troops and squadrons as divisional, corps, or field force troops; in some cases gunships from air cavalry were used for armed escort and scout helicopters for staff visits. These practices, while a boon to senior headquarters, did all too little for tactical commanders at brigade level and below. Only in the 11th Armored Cavalry were air and ground cavalry integrated into a single operational team under a brigade level commander. Colonel George S. Patton once said that the operation of the 11th Cavalry when he was in command really depended on the eyes of those nine warrant officers riding as scouts in his regimental air cavalry troop. The employment of integrated air and ground cavalry must be fully developed and expanded if we are to realize the full potential of the new reconnaissance team.

Third among the lessons taught by Vietnam is what can be done in area and route security, especially in an area traditionally considered the rear. In Vietnam of course there was no rear area; the

M48A3 Tank Explodes a 750-pound Bomb Set Up as a Mine. *Turret was hurled from tank, which was blown out of its tracks.*

enemy was all around. Such a situation could be encountered in a fast-moving war. Usually the U.S. Army has used armored cavalry and other armored units for rear area security.

In the II Corps Tactical Zone for most of the war the 1st Battalion, 69th Armor, and 1st Squadron, 10th Cavalry, acted as reaction forces. At one point in 1970 the 11th Cavalry in the III Corps Zone was daily clearing mines from and providing security for almost 100 miles of logistical resupply routes and farm-to-market roads. While for many reasons armored units are good at this work, the practice can be, and indeed was in Vietnam, a considerable drain on combat forces capable of accomplishing much more for their commander than clearing roads and protecting logistical units.

With limited combat forces at our disposal, it would seem far better to equip and train logistical units to protect themselves, and to furnish area security by providing military police or other units mounted in armored cars and firing weapons designed for the form of enemy resistance they can expect to encounter. In Vietnam some military police units were equipped with armored cars for this purpose, but the system was never widely used. Province chiefs late in the war had their own provincial reconnaissance units mounted in armored cars, and these essentially performed rear area and route security operations. From the standpoint of returns for manpower and equipment invested, it was a far more cost effective operation than assigning a tank, a mechanized infantry, or an armored cavalry unit the same task. The concept of furnishing protection for rear areas and resupply routes in part with the units

ENGINEERS CLEAR TRAIL OF MINES IN CAMBODIA

stationed in the area and in part with a military police type of unit equipped for this purpose needs full exploration and development.

Fourth among the lessons the Vietnam War offers us is the proof that we still need to find better ways of dealing with land mines. Because of the nature of the war, the enemy was able to do great damage with random mines, some of which were relatively simple. Historically, antiarmor land mines have been a persistent and vexing problem for which no really satisfactory solution has ever been found. Our failure to solve the problem of mines laid in patterns has been aggravated by our similar failure to cope with random mining tactics. We must capitalize, therefore, on the experience the U.S. Army gained in dealing with enemy random mining techniques in Vietnam. We must work out a system for using random mines against armor ourselves. And, finally, since random mining can be used against us again, we should develop equipment for swift search and elimination of such land mines. Since World War II almost nothing has been done in this field. The mine rollers sent to Vietnam were not as effective as some 1945 equipment.

The body of experience in logistical support for armored units in Vietnam has useful lessons in maintenance, supply, and battlefield recovery. Maintenance units tended to operate well to the rear. Considerable pressure was required in many cases to persuade them that they could and should operate teams as far forward as squadron and battalion, making repairs on the site at company, troop, and battery level. The alternative was a long haul of damaged equipment back to a maintenance camp and a long haul of repaired equipment back to the unit—a very expensive procedure. At one point, the 11th Armored Cavalry was hauling its damaged Sheridans nearly 150 kilometers round trip. The fact that such a situation existed calls for some reexamination of traditional direct and general support relationships. Some way must be found to provide better security for rear area support units and the routes to and from their customers. Otherwise the customer pays the price to secure the rear and the routes. This cannot go on. Perhaps we have too many intermediate levels of maintenance to operate effectively any longer. Whether or not this is true, we need to find out.

U.S. Army logistical policy calls for area support by maintenance and supply units. In short, support units provided maintenance and supply so long as the using unit was in the geographic area the supporting unit was assigned to support. When the unit moved to another area, its support then came from a unit charged with support in the new area. The problem is that the parts supply system functions on equipment densities and spare parts usage rates. There is not now, and never has been, any satisfactory way to transfer along with the customer unit its experience factors and supply stocks, built up in the supporting unit on the basis of the customer usage factors. The result—in the eyes of the using unit—was that support broke down completely when the unit moved to a new area. At best the spare parts supply system was capable of filling no more than 50 to 60 percent of unit demands; the remaining 40 to 50 percent were filled by cannibalization of machines no longer useful in combat and by going outside the normal supply system—in other words by scrounging parts. On a battlefield in mobile warfare, even this system breaks down. Armored units must have immediately available direct support maintenance and supply as well as adequate backup. In any event, the maintenance and supply methods are in need of close scrutiny and change.

The supply vehicle fleet provided for American armored units in Vietnam was generally unsuited to its tasks. In a country with few and poor secondary roads, it was necessary to replace wheeled cargo carriers with full-tracked cargo vehicles—M548's. These vehicles were essential to the operations of armored units in wide areas along the borders; the Cambodian expedition, for example, could not have been undertaken without them. They were not, however, provided in sufficient numbers in Vietnam and their maintenance reliability was suspect.

M578 LIGHT RECOVERY VEHICLE WORKS ON SHERIDAN

Armored units have always been plagued with the problems of whether their supply fleet should be capable of operating on roads or cross-country, or both. In an attempt to design vehicles that would do both, neither capability was provided satisfactorily. In forward areas, especially in countries with limited road nets, tracked re-supply vehicles at unit level are essential. On the other hand, somewhere there must be a vehicle fleet which can move large volumes of supplies quickly over roads—even if those roads are secondary by some standards. This is primarily an organizational and equipment problem. However, the M548 was the last of its kind; therefore the U.S. Army needs to look seriously at the tracked cross-country resupply capability in forward areas, as well as the long-haul fleet that backs it up.

Recovery of damaged or inoperative vehicles has always been a difficult problem for armored units, and so it was in Vietnam. In respect to both number and reliability of vehicles used, the recovery system was inadequate. The M578 and the M88—the bulk of the recovery fleet—were in short supply for cavalry units, and the M578 was not well designed for its job. The 11th Armored Cavalry attacked into Cambodia with its organic recovery fleet bolstered by almost a dozen M88's borrowed for the occasion out of depot stocks. For almost two weeks, regimental maintenance operations lived on the guts and staying power of these vehicles and their crews.

Recovery of damaged armored vehicles is both an organizational and a doctrinal problem. Normally unit recovery equipment evacuated vehicles to a collecting point where the vehicles were picked up by support units that moved along behind the forward elements. With support units immobilized far to the rear, the burden of battlefield recovery fell to the fighting units—a situation quite likely to recur on a battlefield in the future. There is, therefore, a need for better recovery equipment and more of it at unit level, and a close look at how the Army intends to recover and evacuate disabled vehicles in future wars.

Much useful experience was gained in the Vietnam War. We have seen that the combined arms team is essential and that fighting with troops mounted is advantageous. It is also plain that the American advisers to the South Vietnamese Army were important but that their preparation for the tasks that confronted them was poor. All these experiences and many more must be carefully analyzed.

As we look to the future it is essential not only that we know the lessons of Vietnam, but that we understand them as well. Understanding them, in their correct context, and relating that to the future will take more time and space than we have had available for this monograph. But it must be done. We can no more turn our backs on our experiences in Vietnam than we can take those experiences, relate them directly to our next battlefield, and so in the end get ready to fight better the war we have just left behind. The wisdom to learn from experience, without merely getting better prepared to relive that experience, is not easily won. But win it we must. We owe it to ourselves and our country. More however, we owe it to the brave men who went, helped us learn the lessons, and paid the price of learning. They left us a large legacy—larger perhaps than we deserve.

Appendix A

Commanders of Cavalry, Armor, and Mechanized Infantry Units

The list that follows includes all commanders of armored units of the U.S. Army in Vietnam at battalion and squadron level and above and the organization to which the armored unit was assigned or attached. When an armored unit was permanently attached or assigned as an independent unit below that level, as in the case of a separate cavalry troop attached to an infantry brigade, the commanders have also been included.

Ground Cavalry Units

1st Squadron, 1st Cavalry [1] (23d Infantry Division—Americal)
Lt. Col Richard Harrington, August 1967–January 1968; Lt. Col. Walter C. Cousland, January–July 1968; Lt. Col. Richard D. Lawrence, July–November 1968; Lt. Col. Philip L. Bolte, November 1968–May 1969; Lt. Col. John H. Dure, May–November 1969; Lt. Col. Richard G. Graves, December 1969–July 1970; Lt. Col. Crosbie E. Saint, July-October 1970; Lt. Col. Sheldon J. Burnett, October 1970–March 1971; Lt. Col. Gene L. Breeding, March–July 1971; Lt. Col. Richard E. Lorix, July 1971–April 1972.

Troop E, 1st Cavalry (11th Infantry Brigade, 23d Infantry Division—Americal)
Capt. Terry L. Alexander, October 1967–May 1968; Capt. James R. Oley, May–November 1968; Capt. Kenneth R. Breeden, November 1968–May 1969; 1st Lt. Tony E. Varda, May–June 1969; Capt. Joe F. Gafle, June 1969–March 1970; Capt. Robert P. Seng, March–June 1970; Capt. Kenneth P. Lord III, July–September 1970; Capt. Richard A. Williams, September–November 1970; Capt. James L. Wilson, November 1970–September 1971; Capt. Michael V. Foster, September–October 1971.

2d Squadron, 1st Cavalry [2] (4th Infantry Division)
Lt. Col. Joseph M. Gay, Jr., August 1967–January 1968; Lt. Col. Charles P. Graham, January–June 1968; Lt. Col. Donald W. Moreau,

[1] Although attached to the 23d Infantry Division, this unit remained assigned to the 1st Armored Division, and troops who served with it may wear an armored division patch on the right shoulder of their uniforms. They and the men of the 2d Squadron (see next fn.) were the first to achieve this distinction since World War II.

[2] Although attached to the 4th Infantry Division, this unit remained assigned to the 2d Armored Division, and troops who served with it may wear the 2d Armored Division patch on the right shoulder.

July–December 1968; Lt. Col. Richard A. Miller, December 1968–June 1969; Lt. Col. John M. Fairey, June–December 1969; Lt. Col. Robert G. Bond, December 1969–June 1970; Lt. Col. Landon P. Whitelaw, June–October 1970.

1st Squadron, 4th Cavalry (1st Infantry Division)
Lt. Col. Paul M. Fisher, October 1965–April 1966; Lt. Col. Leonard L. Lewane, April–December 1966; Lt. Col. Thomas W. Fife, December 1966–May 1967; Lt. Col. John W. Seigle, May 1967–February 1968; Lt. Col. Thomas B. Tyree, February–July 1968; Lt. Col. John C. Faith, July 1968–January 1969; Lt. Col. William C. Haponski, January–July 1969; Lt. Col. John T. Murchinson, July–December 1969; Lt. Col. Frederic J. Brown, January–April 1970.

3d Squadron, 4th Cavalry (25th Infantry Division)
Lt. Col. John R. Hendry, February–November 1966; Lt. Col. George S. Webb, November 1966–May 1967; Maj. George N. Stenehjem, May–June 1967; Lt. Col. John M. Shea, June–November 1967; Lt. Col. Glenn K. Otis, December 1967–May 1968; Maj. James G. Jordan, May–June 1968; Lt. Col. Clemens A. Riley, June–November 1968; Lt. Col. Robert S. McGowan, November 1968–August 1969; Lt. Col. Joseph R. Paluh, August–November 1969; Lt. Col. Corwin A. Mitchell, November 1969–May 1970; Lt. Col. Noel D. Knotts, May–October 1970; Lt. Col. Dan L. Drury, October 1970.

3d Squadron, 5th Cavalry (9th Infantry Division)
Lt. Col. Sidney S. Haszard, January–July 1967; Lt. Col. Howard R. Fuller, Jr., July 1967–January 1968; Lt. Col. Hugh J. Bartley, January–August 1968; Lt. Col. Angelo Grills, August–December 1968; Lt. Col. Thomas E. Carpenter, December 1968–June 1969; Lt. Col. Joseph L. Hadaway, July 1969–January 1970; Lt. Col. William N. Bradberry, Jr., January–July 1970; Lt. Col. Harold R. Page, July–December 1970; Lt. Col. Robert B. Osborn, January–October 1971.

1st Squadron, 10th Cavalry (4th Infantry Division)
Lt. Col. Wallace H. Nutting, September 1966–February 1967; Lt. Col. Thomas F. Cole, February–August 1967; Lt. Col. Charles K. Heiden, August 1967–February 1968; Lt. Col. John F. Brownfield, February–July 1968; Lt. Col. Robert W. Noce, July 1968–January 1969; Lt. Col. Roderick D. Renick, Jr., January–June 1969; Lt. Col. Clyde H. Patterson, June–July 1969; Lt. Col. James T. Bramlett, July–September 1969; Lt. Col. William J. Moran, September 1969–March 1970; Lt. Col. Servetus T. Ashworth III, March–August 1970; Lt. Col. John Mason, August 1970–February 1971; Lt. Col. Peter C. Hains, February–August 1971; Maj. John D. George, Jr., August–September 1971; Lt. Col. John W. Hudachek, September–November 1971.

11th Armored Cavalry (II Field Force, Vietnam)
Col. William W. Cobb, September 1966–May 1967; Col. Roy W. Farley, May–December 1967; Col. Jack MacFarlane, December 1967–March 1968; Col. Leonard D. Holder, March 1968; Col. Charles R. Gorder, March–July 1968; Col. George S. Patton, July 1968–April 1969; Col. James H. Leach, April–December 1969; Col. Donn A. Starry, December 1969–June 1970; Col. John L. Gerrity, June–December 1970; Col. Wallace H. Nutting, December 1970–March 1971.

1st Squadron, 11th Armored Cavalry
Lt. Col. Martin D. Howell, September 1966–September 1967; Lt. Col. James H. Holt, September 1967–January 1968; Lt. Col. Jack W. Nielsen, January–August 1968; Lt. Col. Briggs H. Jones, August–November 1968; Lt. Col. Merritte W. Ireland, November 1968–April 1969; Maj. John C. Bahnsen, Jr., April–September 1969; Lt. Col. John M. Norton, September 1969–January 1970; Lt. Col. James B. Reed, January–September 1970; Lt. Col. Donald A. McKnight, September 1970–March 1971.

2d Squadron, 11th Armored Cavalry
Lt. Col. Kibbey M. Horne, September 1966–February 1967; Lt. Col. Benjamin F. Harmon III, February–August 1967; Lt. Col. Garland McSpadden, August 1967–April 1968; Lt. Col. John P. Prillaman, April–September 1968; Lt. Col. Lee D. Duke, October 1968–April 1969; Lt. Col. James H. Aarestad, April–September 1969; Lt. Col. Gail L. Brookshire, September 1969–June 1970; Lt. Col. Richard L. Coffman, June–July 1970; Lt. Col. John L. Ballantyne, August 1970–June 1971; Lt. Col. William M. Stokes III, June 1971–April 1972.

3d Squadron, 11th Armored Cavalry
Lt. Col. Palmer A. Peterson, September–December 1966; Lt. Col. Arthur F. Cochran, December 1966–July 1967; Lt. Col. Hillman Dickinson, July 1967–January 1968; Lt. Col. Neal Creighton, January–July 1968; Lt. Col. John W. McEnery, July 1968–June 1969; Lt. Col. David K. Doyle, June–December 1969; Lt. Col. George C. Hoffmaster, Jr., December 1969–March 1970; Lt. Col. Bobby F. Griffin, March–October 1970; Lt. Col. Frank E. Varljen, October 1970–March 1971.

Troop A, 4th Squadron, 12th Cavalry (1st Brigade, 5th Infantry Division–Mechanized)
Capt. Errol D. Alexander, July–October 1968; Capt. Kenneth G. Carlson, October 1968–March 1969; Capt. Larry R. Robinson, March–September 1969; Capt. William C. Kaufman, September–November 1969; Capt. Matthias A. Spruill, November 1969–February 1970; Capt. John L. B. Smith, Jr., February–May 1970; Capt.

Robert R. Richards, May–October 1970; Capt. Woodrow W. Waldrop, October 1970–April 1971; Capt. Edward E. Helton, April–October 1971.

Troop B, 1st Squadron, 17th Cavalry (82d Airborne Division)
Capt. Dennis P. Malcor, February–May 1968; Capt. William R. Porter, May–November 1968; Capt. Rae W. Dehncke, November 1968–June 1969; Capt. John R. Cushing, June 1969–February 1970.

2d Squadron, 17th Cavalry[3] (101st Airborne Division)
Lt. Col. Julius W. Becton, Jr., December 1967–June 1968; Lt. Col. Egbert B. Clark III, June–December 1968.

Troop A, 2d Squadron, 17th Cavalry[4] (101st Airborne Division)
Capt. William R. Wilson, July 1965–January 1966; Capt. George A. Hamilton, January–June 1966; 1st Lt. Doye W. Adams, June–July 1966; Capt. Richard R. Maglin, July–November 1966; Capt. Dale N. Wagner, November 1966–May 1967; Capt. James R. Harding, May–December 1967.

Troop D, 17th Cavalry[5] (199th Infantry Brigade (Light))
Capt. Michael C. Small, December 1966–November 1967; Capt. Alroy Wahl, November 1967–May 1968; Capt. Keith J. Phillips, May–October 1968; Capt. William B. Garber, October 1968–March 1969; Capt. Paul J. Peterson, March–August 1969; Capt. Michael A. Doyle, August 1969–April 1970; Capt. David F. Rathje, May–October 1970.

Troop E, 17th Cavalry (173d Airborne Brigade)
Capt. David G. Moore, May 1965–April 1966; Maj. Bryan J. Sutton, April 1966–April 1967; Maj. David E. Person, April–September 1967; Capt. Corless W. Mitchell, September 1967–April 1968; Capt. James Ording, April–November 1968; Capt. William J. Hurley, November 1968–April 1969; Capt. Mason B. Sesler, Jr., May–August 1969; Capt. Frederick N. Suttle, August–October 1969; Capt. John M. Weaver, October 1969–April 1970; Capt. Alfred W. Baker, April–June 1970; Capt. John C. Johnston, July–August 1970; Capt. David F. Cunningham, September–December 1970; Capt. Rand L. Allen, December 1970–February 1971; Capt. Edward A. Boles, February 1971–August 1971.

---

[3] During the period December 1968 to June 1969, this squadron was converted to an air cavalry squadron as the 101st Airborne Division converted to an airmobile division.

[4] This troop deployed with the 1st Brigade, 101st Airborne Division, in July 1965 and rejoined the 2d Squadron, 17th Cavalry, when the squadron deployed in December 1967.

[5] Troop D was inactivated on 12 October 1970, but reactivated April 1972 as an air cavalry troop with equipment and troops from Troop D, 1st Squadron, 1st Cavalry.

Troop F, 17th Cavalry (196th Infantry Brigade, 23d Infantry Division—Americal)

Capt. Clyde W. Roan, September 1965–December 1966; Capt. Ronald A. Lane, January–August 1967; Capt. Richard C. Stubbs, August 1967–February 1968; Capt. Richard N. Davis, February–May 1968; Capt. James E. Owens, Jr., May 1968–May 1969; Capt. Alfred I. Stober, June–September 1969; Capt. Klien S. Harrison, September 1969–January 1970; Capt. Roscoe Cartwright, February–September 1970; Capt. Timothy A. Fisher, October 1970–January 1971; Capt. Chester W. Fetner, January–April 1971; 1st Lt. William H. Clarke III, April–May 1971; Capt. George A. Lynn, Jr., May–July 1971; Capt. Charles F. Brower IV, July 1971–April 1972.

Troop H, 17th Cavalry[6] (198th Infantry Brigade, 23d Infantry Division)

Capt. Walter E. Reasor, October 1967–July 1968; Capt. Alfred R. Chioffe, August 1968–March 1969; Capt. James W. Watts, March–May 1969; Capt. John R. Dethorn, May–November 1969; Capt. Max K. Natzel, November 1969–February 1970; 1st Lt. Daniel E. Gooding, March–June 1970; Capt. Oliver L. Croom, Jr., June–November 1970; Capt. Ronald L. Barnable, November 1970–February 1971; Capt. Paul R. Davis, February–April 1971; Capt. Stanley F. Cherrie, April–May 1971; Capt. Randall F. Jarmon, May–August 1971; Capt. Daniel J. Cox, August–October 1971.

Air Cavalry Units

Troop D, 1st Squadron, 1st Cavalry[7] (101st Airborne Division)

Maj. Connie D. Eady, July–November 1968; Maj. Robert E. Wolfe, November 1968–July 1969.

7th Squadron, 1st Cavalry (164th Aviation Group, 1st Aviation Brigade)

Lt. Col. Charles E. Canedy, February–May 1968; Lt. Col. Robert W. Mills, May–November 1968; Lt. Col. George R. Crook, November 1968–May 1969; Lt. Col. George E. Derrick, May–October 1969; Lt. Col. Ronald T. Walker, October 1969–April 1970; Lt. Col. John W. Woodmansee, April–October 1970; Lt. Col. James D. Marett, October–November 1970; Lt. Col. George W. Shallcross, November 1970–September 1971; Lt. Col. Francis B. Martin, September 1971–April 1972.

---

[6] This troop was inactivated in October 1971 but reactivated in April 1972 as an air cavalry troop with troops and equipment from the 7th Squadron, 17th Cavalry (Air).

[7] Troop D was deployed in July 1968 and attached to the 101st Airborne Division until 1969, when it rejoined its parent unit, the 1st Squadron, 1st Cavalry.

Troop F, 4th Cavalry [8] (12th Aviation Group, 1st Aviation Brigade)
Maj. Thomas R. Hamilton, February–June 1971; Capt. Richard A. Bell, June–July 1971; Capt. Rodolfo Gutierrez, July-September 1971; Maj. Billy F. Hatch, September 1971–February 1972; Maj. John J. Spencer, Jr., February–September 1972; Maj. Kermit E. Larson, Jr., September 1972–February 1973.

Troop F, 8th Cavalry (23d Infantry Division—Americal)
Maj. Richard H. Merritt, October 1967–April 1968; Maj. Harold J. Earwood, Jr., April–September 1968; Maj. Lawrence Zittrain, September 1968–February 1969; Maj. Jessie W. Watson, Jr., February–August 1969; Maj. Charles E. Ivey, September 1969–March 1970; Maj. Paul R. Stalker, March–October 1970; Maj. George W. Sibert, October 1970–April 1971; Maj. Alfred S. Rushatz, April 1971–January 1972; Maj. William J. Head, January–March 1972; Maj. John P. Kennedy, March–October 1972; Maj. William D. Dantzler, October 1972–February 1973.

1st Squadron, 9th Cavalry (1st Cavalry Division—Airmobile)
Lt. Col. John B. Stockton, September–December 1965; Lt. Col. Robert M. Shoemaker, December 1965–May 1966; Lt. Col. James C. Smith, May–November 1966; Lt. Col. A. T. Pumphrey, November 1966–April 1967; Lt. Col. Robert H. Nevins, April–December 1967; Lt. Col. Richard W. Diller, December 1967–July 1968; Lt. Col. William C. Rousse, July 1968–January 1969; Lt. Col. James M. Peterson, January–June 1969; Lt. Col. James W. Booth, June–August 1969; Lt. Col. Edward B. Covington, August–September 1969; Lt. Col. James W. Booth, September–December 1969; Lt. Col. Clark A. Burnett, December 1969–August 1970; Lt. Col. Robert H. Nevins, August 1970–January 1971; Lt. Col. Carl M. Putnam, January–June 1971.

Troop F, 9th Cavalry [9] (12th Aviation Group, 1st Aviation Brigade)
Capt. John E. Shields, June–September 1971; Maj. Coleman J. McDevitt, October 1971–April 1972; Maj. George P. Hewlett, May 1972–February 1973.

Troop H, 10th Cavalry [10] (17th Aviation Group, 1st Aviation Brigade)
Maj. Edward M. Brown, April–July 1972; Maj. Sidney F. Lyons, Jr., July 1972–March 1973.

---

[8] Troop F was redesignated from Troop D, 3d Squadron, 4th Cavalry, when the 4th Cavalry redeployed.

[9] Troop F was formed in June 1971 as Troop H, 16th Cavalry, a designation never approved by the Department of the Army. It was renamed Troop F, 9th Cavalry in May 1972.

[10] Activated in April 1972 with troops and equipment from the 7th Squadron, 17th Cavalry (Air).

Troop C, 16th Cavalry (164th Aviation Group, 1st Aviation Brigade)
Maj. Robert L. Phillips, March–July 1970; Maj. Donald M. Frierson, July–December 1970; Maj. Leslie J. Valouche, December 1970–June 1971; Maj. John R. Crist, June–September 1971; Maj. Sanderson A. Woods, September 1971–April 1972; Maj. Thomas H. Craft, April–September 1972; Maj. Donald J. Fritsche, September 1972–February 1973.

2d Squadron, 17th Cavalry [11] (101st Airborne Division, Airmobile)
Lt. Col. William W. De Loach, December 1968–September 1969; Lt. Col. Lavere W. Bindrup, September 1969–March 1970; Lt. Col. Robert F. Molinelli, March 1970–March 1971; Lt. Col. Archie A. Rider, March–October 1971; Lt. Col. Alman I. Butler, October 1971–February 1972.

3d Squadron, 17th Cavalry (12th Aviation Group, 1st Aviation Brigade)
Lt. Col. Christopher B. Sinclair, Jr., October 1967–February 1968; Lt. Col. William W. Brannon, Jr., March–July 1968; Lt. Col. John H. Phillips, July 1968–January 1969; Lt. Col. John B. Fitch, January–September 1969; Lt. Col. Robert A. Arnet, September 1969–February 1970; Lt. Col. Gordon T. Carey, February–September 1970; Lt. Col. Billie G. Williams, September 1970–March 1971; Lt. Col. Carl J. Haaland, March–October 1971; Lt. Col. John E. Dugan, October 1971–March 1972.

7th Squadron, 17th Cavalry (17th Aviation Group, 1st Aviation Brigade)
Lt. Col. Lawrence H. Johnson, October 1967–March 1968; Lt. Col. Stephen F. Cameron, March–September 1968; Lt. Col. Robert M. Reuter, September 1968–March 1969; Lt. Col. Calvin R. Bean, March–September 1969; Lt. Col. George S. Murry, September 1969–March 1970; Lt. Col. Rudolph B. DeFrance, March–September 1970; Lt. Col. Ernest A. Smart, September 1970–September 1971; Lt. Col. Jack W. Anderson, Jr., October 1971–July 1972.

Troop D, 17th Cavalry [12] (11th Aviation Group, 1st Aviation Brigade)
Capt. James W. Bryant, Jr., April–June 1972; Maj. Robert Fairweather, June–September 1972; Maj. Wilbert W. Sorenson, September 1972–February 1973.

Troop H, 17th Cavalry [13] (17th Aviation Group, 1st Aviation Brigade)

---

11 Originally deployed in December 1967 as a ground cavalry squadron, this unit was converted into an air cavalry squadron from December 1968 to June 1969.

12 Troop D was initially deployed as a ground cavalry troop in December 1966 and inactivated in October 1970. Reactivated April 1972 as an air cavalry troop.

13 Troop H was initially deployed as a ground cavalry troop in October 1967 and inactivated in October 1971. It was reactivated with troops and equipment from the 7th Squadron, 17th Cavalry (Air), as an air cavalry troop.

Maj. James M. Gibbs, April–September 1972; Maj. Ronald M. Fishburn, September 1972–February 1973.

Mechanized Infantry Units

2d Battalion, 2d Infantry[14] (1st Infantry Division)
Lt. Col. Edward J. Collins, December 1966–May 1967; Lt. Col. John D. Pelton, May–September 1967; Lt. Col. Henry L. Davisson, September 1967–July 1968; Lt. Col. George D. Greer, July–December 1968; Lt. Col. James A. Michienzi, December 1968–June 1969; Lt. Col. Newell E. Vinson, June–December 1969; Lt. Col. Lee D. Brown, December 1969–April 1970.

1st Brigade, 5th Infantry Division–Mechanized (XXIV Corps)
Col. Richard J. Glikes, July–October 1968; Col. James M. Gibson, October 1968–June 1969; Col. John L. Osteen, Jr., June–December 1969; Brig. Gen. William A. Burke, January–July 1970; Brig. Gen. John G. Hill, Jr., July 1970–May 1971; Brig. Gen. Harold H. Dunwoody, May–July 1971.

1st Battalion, 5th Infantry (25th Infantry Division)
Lt. Col. Thomas U. Greer, January–August 1966; Lt. Col. Victor F. Diaz, August 1966–January 1967; Lt. Col. Richard C. Rogers, January–May 1967; Lt. Col. Chandler Goodnow, May–October 1967; Lt. Col. Fremont B. Hodson, October–December 1967; Maj. Ralph K. Hook, December 1967–January 1968; Lt. Col. Henry B. Murphy, Jr., January–February 1968; Lt. Col. Thomas C. Lodge, February–June 1968; Lt. Col. Andrew H. Anderson, June–October 1968; Lt. Col. William E. Klein, October 1968–June 1969, Lt. Col. Robert A. Kurek, June–October 1969; Lt. Col. Frederick C. Delisle, October 1969–January 1970; Lt. Col. Ted G. Westerman, January–July 1970; Lt. Col. Oliver B. Combs, Jr., July–November 1970; Lt. Col. Patrick J. Moore, November 1970–April 1971.

2d Battalion, 8th Infantry[15] (4th Infantry Division)
Lt. Col. Gordon J. Duquemin, November 1966–May 1967; Lt. Col. John P. Berres, June–December 1967; Lt. Col. John D. Edgerton, December 1967–September 1968; Lt. Col. David P. Thoreson, September 1968–February 1969; Lt. Col. William S. DeCamp, February–July 1969; Lt. Col. Alfred G. Sapp, July–September 1969; Lt. Col. Charles E. Thomann, September 1969–January 1970; Lt. Col.

[14] Originally deployed in October 1965 as a dismounted infantry battalion, this unit was mechanized in January 1967.

[15] Originally deployed in August 1966 as a dismounted infantry battalion, this unit was mechanized in March 1967.

Robert J. Sunell, February–July 1970; Lt. Col. John C. Gazlay, July–October 1970.

1st Battalion, 16th Infantry [16] (1st Infantry Division)
Lt. Col. Robert C. Lewis, October–November 1968; Lt. Col. Donald C. Shuffstall, November 1968–April 1969; Lt. Col. Kenneth G. Cassels, April–September 1969; Lt. Col. David C. Martin, October 1969–April 1970.

2d Battalion, 22d Infantry (25th Infantry Division)
Lt. Col. Richard W. Clark, October 1966–February 1967; Lt. Col. Ralph W. Julian, February-September 1967; Lt. Col. Awbrey G. Norris, September 1967–February 1968; Lt. Col. King J. Coffman, February–August 1968; Lt. Col. James A. Damon, August–December 1968; Lt. Col. Ralph M. Cline, Jr., December 1968–May 1969; Lt. Col. John C. Eitel, May–August 1969; Lt. Col. Bruce F. Williams, August–November 1969; Lt. Col. John R. Parker, November 1969–May 1970; Lt. Col. Nathan C. Vail, May–November 1970.

4th Battalion, 23d Infantry [17] (25th Infantry Division)
Lt. Col. Walworth F. Williams, December 1966–June 1967; Lt. Col. Thomas A. Ware, Jr., June–December 1967; Lt. Col. Avery S. Fullerton, December 1967–May 1968; Lt. Col. Clifford C. Neilson, May–November 1968; Lt. Col. Albert C. Butler, December 1968–March 1969; Lt. Col. George E. Taylor, March–September 1969; Lt. Col. James E. Coggins, September 1969–March 1970; Maj. Frederick S. Stanley, March–April 1970; Lt. Col. Edward M. Bradford, April–November 1970; Maj. Jerry D. Blackwood, November–December 1970.

2d Battalion, 47th Infantry (9th Infantry Division)
Lt. Col. William B. Cronin, January–April 1967; Lt. Col. Arthur D. Moreland, April–December 1967; Lt. Col. John B. Tower, January–June 1968; Lt. Col. Frederick F. Van Deusen, June–July 1968; Lt. Col. James L. Scovel, July 1968–January 1969; Lt. Col. Douglas S. Smith, January–July 1969; Lt. Col. James W. Rowe, July–January 1970; Lt. Col. John H. Claybrook, January–July 1970; Lt. Col. Gary C. Williams, July–October 1970.

1st Battalion, 50th Infantry (I Field Force, Vietnam)
Lt. Col. Albert L. Hutson, Jr., September 1967–January 1968; Lt. Col. Cheney L. Bertholf, Jr., February–July 1968; Lt. Col. John B.

---

16 Deployed in December 1966 as the 5th Battalion, 60th Infantry, this unit was reassigned and redesignated as the 1st Battalion, 16th Infantry.

17 This unit was originally deployed in April 1966 as a dismounted infantry battalion but was mechanized in January 1967.

Carter, July 1968–January 1969; Lt. Col. James R. Woodall, January–June 1969; Maj. Oren R. Culpepper, June–July 1969; Lt. Col. John M. Gilbert, July 1969–January 1970; Lt. Col. Robert H. Luck, January–June 1970; Lt. Col. Richard D. Hooker, June–December 1970.

5th Battalion, 60th Infantry [18] (9th Infantry Division)

Lt. Col. Lucian K. Truscott III, December 1966–March 1967; Lt. Col. Allan S. Flynn, March–August 1967; Lt. Col. William B. Steele, August 1967–March 1968; Lt. Col. Eric F. Antila, March–September 1968.

1st Battalion, 61st Infantry (1st Brigade, 5th Infantry Division—Mechanized)

Lt. Col. Bernard D. Wheeler, July–December 1968; Lt. Col. David E. Hartigan, January–July 1969; Lt. Col. John W. Swarcu, Jr., July 1969–June 1970; Lt. Col. Richard A. Scholtes, June 1970–February 1971; Lt. Col. Arnold S. Stallman, February–July 1971; Lt. Col. Robert M. Brumback, July 1971.

Tank Units

Company D, 16th Armor (173d Airborne Brigade)

Capt. Josef C. Jordan, Jr., May–September 1965; Capt. John E. Dunlop, Jr., September 1965–April 1966; Capt. Karl F. Nehammer, April–October 1966; Capt. John K. Waters, October 1966–May 1967; Capt. Robert D. Mackey, May–November 1967; Capt. Robert F. Helmick, November 1967–April 1968; Capt. James S. Hicks, April–December 1968; Capt. Ronald A. Bogue, December 1968–July 1969; Capt. John M. Weaver, July–October 1969.

2d Battalion, 34th Armor [19] (25th Infantry Division)

Lt. Col. Raymond L. Stailey, September 1966–August 1967; Lt. Col. Hal B. Rhyne, August 1967–February 1968; Lt. Col. John H. Tipton, February–July 1968; Lt. Col. Theodore E. O'Connor, July 1968–January 1969; Lt. Col. Duane R. Tague, January–June 1969; Lt. Col. Tommie G. Smith, June–November 1969; Lt. Col. William M. Greenberg, December 1969–May 1970; Lt. Col. Birtrun S. Kidwell, May–October 1970.

---

[18] Reassigned to the 1st Infantry Division in September 1968 and redesignated as the 1st Battalion, 16th Infantry.

[19] Originally assigned to and deployed with the 4th Infantry Division, this unit was under control of II Field Forces, Vietnam, and then attached to the 25th Infantry Division.

1st Battalion, 69th Armor [20] (4th Infantry Division)

Lt. Col. Ronald J. Fairfield, Jr., January–September 1966; Lt. Col. Clyde O. Clark, September 1966–March 1967; Lt. Col. Paul S. Williams, Jr., March–September 1967; Lt. Col. William D. Grant, September 1967–February 1968; Lt. Col. Theodore S. Riggs, Jr., February–August 1968; Lt. Col. Stan R. Sheridan, September 1968–January 1969; Lt. Col. Leo M. Brandt, January–May 1969; Maj. George J. Latturner, May–July 1969; Lt. Col. Donald J. Pagel, July–November 1969; Lt. Col. James L. Marini, December 1969–April 1970.

1st Battalion, 77th Armor (1st Brigade, 5th Infantry Division—Mechanized)

Lt. Col. John M. Pickarts, July–December 1968; Lt. Col. Carmelo P. Milia, December 1968–May 1969; Lt. Col. Thomas A. Miller, Jr., June–December 1969; Lt. Col. Niven J. Baird, December 1969–June 1970; Lt. Col. John T. McNamara, July–November 1970; Lt. Col. Richard M. Meyer, November 1970–April 1971; Lt. Col. Robert E. Butler, April–July 1971.

Special Armored Unit

39th Cavalry Platoon (Air Cushion) [21] (9th Infantry Division)

Maj. David G. Moore, May 1968–April 1969; Maj. Edward R. Szeman, April–December 1969; Maj. Duane B. Root, December 1969–June 1970; Maj. Barry F. Graham, July–August 1970; 1st Lt. George E. Rogers, August–September 1970.

---

[20] Originally assigned to and deployed with the 25th Infantry Division, this unit was attached to the 4th Infantry Division.

[21] This unit was formed in Vietnam using three British-built Hovercraft. The unit name, never officially approved by the Department of the Army, was developed from the numerals of the 3d Brigade, 9th Infantry Division, to which the Hovercraft were attached.

Appendix B

Armor Recipients of the Medal of Honor

	Province
Bondsteel, S. Sgt. James L., Co A, 2d Bn (M), 2d Inf, 1st Inf Div	An Loc 24 May 1969
Doane, 1st Lt. Stephen H., Co B, 1st Bn (M), 5th Inf, 25th Inf Div	Hau Nghia 25 March 1969
Fernandez, Sp4c. Daniel, Co C, 1st Bn (M), 5th Inf, 25th Inf Div	Hau Nghia 18 February 1966
Fitzmaurice, Sp4c. Michael J., Trp D, 2d Sqdn (Air), 17th Cav, 101st Airborne Div (Ambl)	Quang Tri 23 March 1971
Fritz, 1st Lt. Harold A., Trp A, 1st Sqdn, 11th Arm Cav Regt	Binh Long 11 January 1969
Johnson, Sp5c. Dwight H., Co B, 1st Bn, 69th Arm, 4th Inf Div	Kontum 15 January 1968
Langhorn, Pfc. Garfield M., Trp C, 7th Sqdn (Air), 17th Cav, 1st Aviation Bde	Pleiku 15 January 1969
LaPointe, Sp4c. Joseph G., Jr., Hq Trp, 2d Sqdn (Air), 17th Cav, 101st Airborne Div (Ambl)	Quang Tin 2 June 1969
Long, Sgt. Donald R., Trp C, 1st Sqdn, 4th Cav, 1st Inf Div	Binh Long 30 June 1966
McKibben, Sgt. Ray, Trp B, 7th Sqdn (Air), 17th Cav, 1st Aviation Bde	Binh Thuan 6 December 1968
Patterson, Sgt. Robert M., Trp B, 2d Sqdn, 17th Cav, 101st Airborne Div	Thua Thien 6 May 1968
Poxon, 1st Lt. Robert L., Trp B, 1st Sqdn (Air), 9th Cav, 1st Cav Div (Ambl)	Tay Ninh 2 June 1969
Skidgel, Sgt. Donald S., Trp D, 1st Sqdn (Air), 9th Cav, 1st Cav Div (Ambl)	Binh Thuan 14 September 1969
Steindam, 1st Lt. Russell A., Trp B, 3d Sqdn, 4th Cav, 25th Inf Div	Tay Ninh 1 February 1970
Taylor, Capt. James A., Trp B, 1st Sqdn, 1st Cav, 23d Inf Div	Quang Nam 9 November 1967
Warren, 1st Lt. John E., Jr., Co C, 2d Bn (M), 22d Inf, 25th Inf Div	Tay Ninh 14 January 1969
Wickham, Cpl. Jerry W., Trp E, 2d Sqdn, 11th Arm Cav Regt	Binh Long 6 January 1968
Yano, Sfc. Rodney J. T., Air Cav Trp, 11th Arm Cav Regt	Bien Hoa 1 January 1969
Young, S. Sgt. Marvin R., Co C, 1st Bn (M), 5th Inf, 25th Inf Div	Hau Nghia 21 August 1968

Glossary

Abn	Airborne
ACAV	Armored cavalry assault vehicle
ACR	Armored cavalry regiment
Aerorifle platoon	A platoon of forty-four enlisted men and one officer organized as light airmobile infantry and transported to mission areas by helicopter. The platoon was specifically organized as an organic element of every air cavalry troop.
Ambl	Airmobile
APC	Armored personnel carrier
Area warfare	That condition of war in which forces orient exclusively on the destruction of enemy forces with no intention to seize and hold a succession of terrain features. Area warfare is nonlineal and multidirectional and has no continuous front or line of demarcation between opposing forces.
ARVN	Army of the Republic of Vietnam
Base area	A section of terrain which contains logistical installations, defensive fortifications, or other physical structures used by the enemy.
Base camp	The location which provides a semipermanent home for tactical organizations.
Canister rounds	Short range antipersonnel projectiles, loaded with submissiles such as flechettes or steel balls. The casing is designed to open just beyond the muzzle of the weapon, dispersing the submissiles.
Clear and secure	A military mission to find and capture or destroy all enemy forces within a specified area and then prevent enemy forces from reentering.
Cloverleaf	A formation used in search operations that allowed a rapid search of a large area.
Cobra	Name given to the attack helicopter AH–IG.
COSVN	Central Office for South Vietnam (Communist headquarters)

CP	Command post
CTZ	Corps tactical zone
Daisy cutter	A 15,000-pound bomb used to create a landing zone in thick jungle.
FFV	Field Force, Vietnam
FSB	Fire support base
Go/No Go	The term applied to the trafficability of tracked vehicles in the Republic of Vietnam.
Gunship	A helicopter armed with air-to-ground armament.
HEAT	High explosive, antitank
Herringbone formation	A formation used by mechanized and armor units in which armored vehicles turn alternately to the sides of the direction of march, placing their main armament and heaviest armor obliquely toward the flanks. The center is left clear to provide freedom of movement within the column and as a haven for vehicles without armor.
Ho Chi Minh Trail	An infiltration route used extensively by the North Vietnamese Army to move troops and supplies from North Vietnam through Laos and Cambodia to all regions of South Vietnam.
LAW	Light antitank weapon
LTL	Interprovincial route
LZ	Landing zone
M	Mechanized
MACOV	Mechanized and Armor Combat Operations in Vietnam, a study.
MACV	Military Assistance Command, Vietnam
Minigun	A six barreled 7.62-mm. machine gun using a system of rapidly rotating barrels which produces a high rate of fire.
MR	Military region
QL	National highway
Recoilless rifle	Weapon in which the rearward movement resulting from firing is essentially eliminated.
Reconnaissance in force	A limited objective operation by a military force to discover and test the enemy's disposition and strength or to develop other intelligence.

Roadrunner operation	Normally an operation with a group of vehicles that travel a road for the purpose of keeping the enemy off balance and making the presence of friendly forces felt among the local populace.
Rome plow	A special bulldozer blade manufactured by the Rome Plow Company and used for cutting brush and small and medium trees.
RPG	Rocket propelled grenade
Sensor	Equipment which detects and indicates terrain configuration, the presence of military targets, and other natural and man-made objects, and activities by means of energy emissions or reflections.
Standdown	Relief from combat operations, usually in preparation for redeployment or for maintenance.
Swift boat	A small modified Navy patrol boat used for close in offshore security and inland waterway patrolling.
Tactical air support	The employment of air force elements in combat in conjunction with ground operations.
Tet	A celebration of the new year based on the lunar calendar. This date fluctuates each year.
TF	Task force
TL	Provincial route
TOC	Tactical operations center
TOW	Tube-launched, optically tracked, wire-guided missile
USMC	United States Marine Corps

Index

A Shau valley: 156, 201, 205
Abrams, General Creighton W.: 35, 38, 138, 139*n*, 143, 153, 164, 199
Advisers: 17, 19–25, 27–31, 33–35, 37, 151, 162–63, 168–69, 177, 180–81, 193, 197, 205, 218, 220–21, 226
Ai Tu: 204, 206
Air Assault Division, 11th: 51
Air cavalry: 51, 57–60, 84–85, 111–12, 116, 130, 134, 193–94, 200, 217, 221
Air cushion vehicles: 142*n*
Air strikes: 52, 202, 215, 217–18
Air supply: 122, 154, 192
Air support of ground operations: 62–63, 67, 69, 78, 97, 100, 108, 122–23, 130, 134, 136, 146, 155, 157, 170–71, 174, 178, 193, 195, 209, 212. *See also* Helicopters
Airborne Brigade, 173d: 55, 65, 112, 161
Airborne Division, 101st: 58, 115*n*, 126, 156
Airborne operations: 27–28, 32
Aircraft: B–52, 202, 215, 217–18; C–47 (Spooky), 97
Airmobile forces: 50, 67, 71, 85, 111, 116, 156
Aloui: 190, 192–96
Ambush actions, enemy: 5, 108–10, 134, 195–96, 198
Ambush/counterambush actions: 34, 50, 59, 66–67, 69–71, 75, 77–78, 81, 94, 101, 125–26, 141, 156, 160
Amphibious operations: 54
Amphibious tracked vehicle, M551 (General Sheridan): 82, 142–45, 187*n*, 224
An Khe: 58
An Loc: 66, 69, 71, 159
An My: 118
Angel's Wing: 167, 169–70
Annamite mountains: 9
Antiaircraft, enemy: 27, 191–92, 201–2
Ap Bac: 25, 27–28, 32, 38
Ap Bac II: 103
Ap Bau Bang: 60, 63, 97, 100
Ap Dong: 123
Ap Tan Thoi: 25
Ap Tau O bridge: 67
Armor: 6, 64, 66–67, 69–71, 75, 77–78, 87; firepower and mobility, 42, 75, 96–97, 100, 103, 105–6, 115–16, 132, 136–37, 139, 144, 155, 164–65, 179, 186*n*, 219; opinion on need for, 7, 51, 55–58, 65, 72, 90; troop withdrawal, 164–65, 199. *See also French Army*
Armor doctrine: 7, 17, 24, 33–34, 41–42, 50–51, 53, 65, 84–87, 90, 95, 220
Armor roles and missions: blocking, 95, 190; border sealing, 156; cordon and search, 74, 103, 140; enclave strategy, 54–55; mobile reserve, 219; reaction force, 50, 53, 67, 95, 115–16, 128, 222; reconnaissance, 97, 103, 161, 221; road and convoy security, 50, 54–55, 58, 74, 81, 95, 103, 106–11, 128, 147, 177, 191, 199, 221–22; screening, 95; search and destroy, 31, 53–55, 64, 91, 94–95, 100, 103, 161, 177; surveillance, 50, 161
Armor tactics: cloverleaf maneuver, 86; herringbone formation, 70, 77–78, 85, 109; pile-on, 145, 147; roadrunner operations, 72, 74, 80, 85, 103; thunder run technique, 71–72, 80, 85
Armored battalions and companies: Company D, 16th Armor, 91; 2d Battalion, 34th Armor, 73, 91, 94, 100, 102, 132, 171, 183; 1st Battalion, 69th Armor, 58, 63, 73, 79, 107, 150–53, 180*n*, 183, 222; 1st Battalion, 77th Armor, 139, 187, 196
Armored cars: 17–18, 25, 30–31, 34–35
Armored cavalry: 50–51, 67, 72, 84–85, 95, 102, 116, 118, 123, 142–43, 145. *See also* Cavalry units
Armored cavalry assault vehicle (ACAV): 73, 84, 89, 107*n*, 143–44; employment of, 95, 99–100, 102, 106, 126, 130, 142, 149, 172*n*; losses, 79, 82, 110, 126, 193, 210. *See also* Armored personnel carrier, M113
Armored Cavalry Regiment, 11th: 58, 72–75, 77–80, 82, 91, 95–96, 108, 126, 142, 182, 184–85, 195–96, 221–22, 224–25; 1st Squadron, 143–44, 154–55, 159–60, 175–76; 2nd Squadron, 154–57, 159–60, 171–74, 176, 199; 3d Squadron, 127, 149, 159, 171–74, 176
Armored divisions: 1st, 107*n*; 2d, 107*n*
Armored personnel carrier, M113: 35, 211*n*, 214; employment of, 21–24, 25, 27–29, 31–32, 34, 36–37, 42–45, 62–63, 66, 72–73, 75, 77, 85–86, 90, 105, 141, 143–44; losses, 63, 70, 79, 119, 134;

modifications to, 34, 38, 40–43, 57, 63–64, 73, 84, 89; problems with, 24, 41–43, 47, 64–65, 79, 81*n*, 82, 97, 103–5. *See also* Armored cavalry assault vehicle (ACAV)
Armored personnel carrier, M113A1: 107*n*, 131*n*, 142*n*
Armored personnel carriers: 10, 45, 51, 64, 112–13, 136*n*, 187
Armored reconnaissance vehicle, M114: 24, 25, 37–38, 57, 72–73
Army Combat Operations in Vietnam study: 89-90
Artillery: 8-inch, 66; 105-mm., 146; 175-mm., 66, 151, 153
Artillery battalions: 5th, 4th Artillery, 139; 7th, 9th Artillery, 99; 2d, 33d Artillery, 60, 62–63; 2d, 77th Artillery, 100
Artillery support of ground operations: 63, 66–67, 69, 78, 80–81, 97, 99–100, 102, 105, 108, 110, 122–23, 125, 130, 132, 135–36, 154–55, 157, 159, 170–71, 174, 178, 193, 195, 209, 215, 217
ATLANTA Operation: 74–75, 78
ATTLEBORO Operation: 74
Australian forces: 55, 112–13
Aviation Group, 164th: 178

Ba, Colonel Ly Tong: 23, 25, 27, 48, 214
Ban Me Thuot: 33, 118, 179
Bao Tri: 64
Bartley, Lieutenant Colonel Hugh J.: 126, 145–47
Base areas and camps, enemy: 64, 91, 96, 131, 138, 148, 155, 179. *See also* Sanctuaries, enemy
Battlefield illumination: 71, 94
Battlefield recovery: 45, 64, 182, 185, 223, 225–26
Battreal, Colonel Raymond R., Jr.: 31, 35, 205
Bau Ba Linh: 157
Bau Bang: 60, 62–63, 96
Bau Long Pond: 60
Bell Helicopter: 215
Bellis, Private First Class Mark: 134
Ben Cat: 91, 118
Ben Hai River: 205, 207
Ben Het: 150–53, 212, 214–15
Bender, Lieutenant Colonel John A.: 102
Berry, Colonel Sidney B., Jr.: 69
Bien Hoa: 74, 123–27, 144, 154
Binh An: 145–47
Binh Dinh Province: 161
Binh Gia: 29
Binh Long Province: 66, 155–56, 159
BINH TAY I–IV Operations: 179
Black Panthers: 194
Boi Loi Woods: 64, 79
Boles, Brigadier General John K., Jr.: 31
Boston: 146
Bowers, Sergeant: 27
Bricker, Captain James W.: 22–23
Bridges: 187; balk, 41, 103; M4T6, 173; vehicle-launched, 34, 172, 203–4
Brookshire, Lieutenant Colonel Grail L.: 157, 159–60, 171
Brown, Major Lloyd J.: 30
Bu Dop: 156
Bulldozers: 91, 94, 148, 187
Bunker, Ellsworth: 164
Burnett, Specialist 4 William D.: 62

Ca Mau Peninsula: 9
Caldwell, Captain John S.: 160
Cam Lo: 205–7, 209
Cambodia: border operations, 59, 66, 67, 115, 151, 156; operations in, 164, 167–74, 177–82, 185, 197–99, 224–25; troop withdrawal, 174, 176–77, 186. *See also* Sanctuaries, enemy
Cambodian forces: 169, 180, 187
Cambodian government: 166–67, 180, 197
Camouflage: 150
Camps: Blackhorse, 74–75, 77, 107–9, 149; Carroll, 205–6, 208; Evans, 212; Hawk Hill, 134
Can Tho: 197
Cannon, 20-mm.: 62, 194
Capital Military District: 129
Cargo carrier, M548: 203–4, 224–25
Cavalry battalions: 1st Battalion, 5th Cavalry, 146; 2d Battalion, 5th Cavalry, 146; 1st Battalion, 8th Cavalry, 59–60, 154–55; 2d Battalion, 12th Cavalry, 59
Cavalry Brigade, 1st, 9th Cavalry: 177*n*
Cavalry Division, 1st: 29, 51, 58, 60, 65–66, 111, 145–46, 154, 156, 159, 170–71, 183
Cavalry Platoon (air cushion vehicle), 39th: 142*n*
Cavalry squadrons: 1st Squadron, 1st Cavalry, 81*n*, 107*n*, 133–36, 187, 200; 2d Squadron, 1st Cavalry, 107, 118; 7th Squadron, 1st Cavalry, 165, 172; 1st Squadron, 4th Cavalry, 56–57, 60–63, 65, 67, 69–71, 91, 95, 107, 123, 129, 143, 180*n*; 3d Squadron, 4th Cavalry, 58, 66, 91, 110–11, 118–19, 122–23, 134, 143–44, 174; 3d Squadron, 5th Cavalry, 99–100, 107–10, 125–26, 145–47, 156,

165, 187, 190; 1st Squadron, 9th Cavalry, 58, 59–60, 66, 111–12, 145, 154, 171, 172*n*, 177*n*; 1st Squadron, 10th Cavalry, 107, 199, 222; 2d Squadron, 17th Cavalry, 115*n*, 190–91, 192*n*, 194, 197*n*; 3d Squadron, 17th Cavalry, 112, 177; 7th Squadron, 17th Cavalry, 112, 150*n*, 200, 212
Cavalry troops: Troop F, 4th Cavalry, 201*n*; Troop F, 8th Cavalry, 136; Troop H, 10th Cavalry, 200; Troop A, 4th Squadron, 12th Cavalry, 139, 187; Troop D, 4th Squadron, 12th Cavalry, 51; Troop C, 16th Cavalry, 200; Troop D, 17th Cavalry, 200; Troop E, 17th Cavalry, 91; Troop F, 17th Cavalry, 200; Troop H, 17th Cavalry, 200
CEDAR FALLS Operation: 91, 94–95, 103, 148
Central Office for South Vietnam (COSVN): 12, 95, 149, 171
Chamberlain, Lieutenant Colonel Edwin W., Jr.: 105
Chu Lai: 54
Chu Pao Pass: 216–17
Chu Pong Mountain: 66
Chup rubber plantation: 176–77
CIRCLE PINES Operation: 64
Civil Guards, Republic of Vietnam: 25, 27, 28
Claymore Corner: 77
Cleveland: 198*n*
Cobb, Colonel William W.: 95
Collins, Lieutenant Colonel Edward J.: 96
Combat Aerial TOW Team, 1st: 215–17
Combined arms team: 50, 60, 63, 84–85, 97, 99–100, 102–3, 113, 130, 132–33, 136, 155, 169, 174, 221, 226
Combustible-case ammunition: 143–44
Command and control: 35, 67, 108, 112. *See also* Republic of Vietnam Army
Cong Ba Ky Canal: 27
Cong Luong Canal: 25, 27
Corps, XXIV: 194, 205
Corps tactical zones: 117*n*; I Corps, 35, 56, 73, 114–16, 133, 139–40, 145, 150, 153, 156, 161–63, 190, 192–95; II Corps, 29, 33, 66, 73, 79, 107, 115–16, 118, 149–50, 161, 179, 222; III Corps, 64, 66, 73, 79, 91, 103, 107, 110, 115, 118, 123–24, 129, 132–33, 147–49, 154, 156, 167–69, 172, 178–79, 222; IV Corps, 24, 27, 43, 115, 127, 162, 168, 172, 175, 178–79. *See also* Military regions
Crow's Nest: 167–68
Cu Chi: 64–65, 73, 118–19, 181
Cu Mong Mountain: 202
Cua Viet River: 206, 210
Cuu Long 15: 33
CUU LONG I–III Operations: 178
CUU LONG 44–02 Operation: 197–98
Da Nang: 52–54, 200
Dak To: 151
Dak To II: 212–13
Dalat: 17
Dam, Colonel Bui Dinh: 27
Dau Tieng: 154
Davis, Specialist 4 Eddie: 152
Demilitarized Zone: 54, 161–62, 187, 205–6, 209
DePuy, Major General William E.: 65, 67
DEWEY CANYON II Operation: 187
Diem, Ngo Dinh: 48–49
Distinguished Service Cross: 62, 123*n*
Dog's Head: 174
Dong Ha: 206–11
Downey, Staff Sergeant John: 52
Duc Co Special Forces Camp: 59
Duc Hoa: 64
Dzu, Lieutenant General Ngo: 212

Edson: 146
EL PASO II Operation: 66
Elliott, Lieutenant Colonel Joe: 101
Engineer battalions: 7th, 187; 169th, 148
Engineer operations: 69, 147–48, 187

Fall, Bernard B.: 4–5
Field Force, II, Vietnam: 65, 73, 91, 95, 115, 123–24, 168
Filhol Plantation: 64
Fire support bases: I, II, III, 95–96; Apple, 125–26; Bastogne, 201–2; Bravo, 196; Elliott, 190; Gold, 100–101; Ruth, 156–57; Vandergrift, 187; Veghel, 202
Fisher, Lieutenant Colonel Paul A.: 57
Fishhook: 159, 171, 172*n*, 174–76
Flamethrower vehicle, M132: 43, 70
Flares: 71, 97, 147, 152
Free World Military Assistance Forces in Vietnam: 31, 91, 114–16, 129, 131, 136–37, 138, 153, 155, 161, 168
French Army: 10, 12; armor experience, 3–6, 8–9, 17, 21, 38, 40, 44, 56, 73, 87, 220; Groupement Mobile 100, 5
Fulton, Colonel William B.: 105

Garcia, Second Lieutenant John: 63
Garner, First Lieutenant Larry: 104
Garretson, Captain Ralph B.: 125–26
Garth, Colonel Marshall: 101
Geneva Accords of 1954: 3, 17

Gia Ray: 75, 77, 107
Giai, General Vu Van: 205-6, 208
"Green dragons": 24*n*
GREEN RIVER Operation: 161
Greer, Lieutenant Colonel Thomas U.: 65
Grenade launcher, 40-mm.: 34, 69
Griffin, Lieutenant Colonel Bobby F.: 171-72, 174
Guiz, Lieutenant Thomas: 133-34, 136
Guns: 20-mm., 41; 40-mm. (twin), 151
Gunshields: 38, 40, 57, 63, 64, 73, 84

Ha Tien: 197
Half-tracks: 3, 17, 18
Hamby, Captain Jerrell F.: 32
Hazzard, Lieutenant Colonel Sidney S.: 99
Hau Nghia Province: 199
Havermale, Sergeant First Class Hugh H.: 151-52
Heiser, Lieutenant General Joseph A. M., Jr.: 182
Helicopters: 3, 50-51, 58, 71; AH-1G (Cobra), 157, 159, 192*n*, 194, CH-47, 69, 135; CH-54, 173; OH-6A, 194; UH-1B, 215; airlift and supply, 25, 27, 32, 122, 154; command and control, 57, 168, 194; fire support of ground operations, 54, 63, 77, 125, 128, 157, 159, 168, 171, 174, 192-95, 198*n*; losses, 27-28. *See also* Air cavalry
Hembree, Specialist 4 Frank: 151-52
Highways: Highway 1, 36-37, 72, 74, 77-78, 111, 119, 122, 124-25, 169-70, 175, 199, 205-7; Highway 2, 74, 109; Highway 3, 197; Highway 4, 95-96, 197-98; Highway 7, 172, 176; Highway 9, 187, 190-92, 196, 206, 209; Highway 13, 60, 62, 66, 67, 69, 71, 73, 96-97, 107, 159, 176-77; Highway 14, 156-57, 159, 214-17; Highway 15, 175-77; Highway 18, 198; Highway 19, 79, 107; Highway 20, 149; Highway 246, 159; Highway 333, 78
Hill, Brigadier General John G., Jr.: 187*n*
Hixson, Chief Warrant Officer Douglas R.: 216
Ho Bo Woods: 64, 79
Ho Chi Minh Trail: 150, 166
Ho Nai: 124
Hoc Mon bridge: 119
Holfmaster, Lieutenant Colonel George C.: 159
Holtom, Captain Stanley E.: 22
Honolulu Conference, 1965: 112
HONOLULU Operation: 64
Howell, Lieutenant Colonel Martin D.: 77-78
Howitzers: 73; M8, 17; M109, 51
Howze, Lieutenant General Hamilton H.: 51
Howze Board: 51
Hue: 54, 114, 116, 128, 183, 201-2, 205
Hughes Aircraft, 215
Hunter, Master Sergeant Andrew: 102

Ia Drang valley: 29, 60
Infantry battalions: 1st Battalion, 2d Infantry, 69; 2d Battalion, 2d Infantry, 60, 80, 95-96, 180*n*; 2d Battalion, 3d Infantry, 125; 1st Battalion, 5th Infantry, 58, 63-65, 79, 91, 132, 174; 2d Battalion, 8th Infantry, 80*n*, 107*n*; 1st Battalion, 11th Infantry, 139; 2d Battalion, 12th Infantry, 100-102; 4th Battalion, 12th Infantry, 125; 1st Battalion, 16th Infantry, 180*n*; 2d Battalion, 22d Infantry, 91, 100; 3d Battalion, 22d Infantry, 100-101, 132, 174; 4th Battalion, 23d Infantry, 80*n*, 132, 118; 2d Battalion, 27th Infantry, 123, 132; 2d Battalion, 47th Infantry, 103, 123-25, 130, 165, 171-72; 3d Battalion, 47th Infantry, 104-5; 1st Battalion, 50th Infantry, 161, 183; 3d Battalion, 60th Infantry, 104-6; 5th Battalion, 60th Infantry, 103-5, 130; 1st Battalion, 61st Infantry, 139, 187
Infantry brigades: 1st Brigade, 5th Division, 82, 139-41, 153, 161-62, 182, 187*n*, 203; 2d Brigade, 9th Division, 103; 3d Brigade, 9th Division, 171; 1st Brigade, 25th Division, 132-33, 174; 2d Brigade, 25th Division, 95, 171; 196th Brigade (Light), 94; 199th Brigade (Light), 123-24
Infantry divisions: 1st Division, 55-57, 60, 63, 65-67, 73, 80*n*, 154, 156, 180*n*; 4th Division, 73, 82, 107, 150*n*; 9th Division, 103, 108, 165; 23d Division, 133, 135; 25th Division, 45, 58, 63-65, 73, 110, 111*n*, 119, 156, 167, 174-75
Intelligence operations and reports: 25, 29, 31, 60, 73, 81, 91, 110, 114, 128, 131, 140, 154, 159-60, 168, 187, 197, 201, 212, 215
Iron Triangle: 91, 148

Johnson, General Harold K.: 55-56, 84
Johnson, Lyndon B.: 58, 164
Joint Research and Test Activity in Vietnam: 31

Jones, Staff Sergeant Jerry W.: 151–52
Jones, Lieutenant Colonel States R., Jr.: 53
Julian, Lieutenant Colonel Ralph W.: 100–101
JUNCTION CITY Operation: 95–97, 100, 102–3, 148
Jungle fighting: 6–7, 60, 64, 84–85

Kampong Cham: 175–77
Kampong Som: 197, 198*n*
Kampong Spean River: 174
Kampong Trabek: 175
Kampong Trach: 174
Katum: 95, 159–60, 176
Kennedy, John F.: 50
Khanh, Major General Nguyen: 49
Khe Sanh: 54, 116, 150, 187, 190
KITTY HAWK Operation: 107–8
Knowles, Brigadier General Richard T.: 94
Kontum: 118, 179, 212, 214–17
Korean forces: 113
Korean War: 6–7, 56, 220
Krek: 174, 176, 201*n*
Kroesen, Major General Frederick J., Jr.: 205

Lai Khe: 60
LAM SON 45–72 Operation: 202
LAM SON 719 Operation: 177, 186–87, 190, 192–93, 195–98, 203
Landing zones: LZ 31, 192–95; Alpha, 195–96; Hope, 195; Ranger North, 191–92; Ranger South, 191–92
Landry, Lieutenant Colonel Walter J.: 17
Lang Vei Special Forces Camp: 150
Lao Dong Party: 12
Laos: 150, 153–54, 156, 177, 186, 190–91, 193, 195, 198. *See also* Sanctuaries, enemy
Lark, Specialist 4 Thomas: 97
LAW, M72: 197, 215
Lawrence, Lieutenant Colonel Raymond D.: 131–36
Leach, Colonel James H.: 155
Lessons learned: 22–24, 28, 33, 67, 71, 108, 111, 163, 170, 220–26
LINCOLN Operation: 66
Listening posts: 50, 97
Loc Ninh: 66, 72, 156, 159–60, 176
Logistical Command, Vietnam, 1st: 182
Lon Nol: 166
Long Binh: 123–27, 181–82
Luat, Colonel Nguyen Trong: 208, 211
Ly, Colonel Nguyen Huu: 207

Machine guns: 97, 102; .30-cal., 41; .50-cal., 22, 23, 27, 38, 42, 62, 69, 73, 142; 7.62-mm., 69, 142; M60, 73
McNamara, Robert S.: 50, 89
Madole, Major James: 31
Mahler, Major Michael D.: 147
Mai, Sergeant Nguyen Xuan: 193
Mai Loc: 206, 208*n*
Maintenance and repair: 21, 30, 44–45, 171*n*, 181–86, 204, 206, 223–24
MAKIKI Operation: 64
Mao Tse-tung: 14, 29
Marshall, Captain Carl B.: 157
Materiel losses: 27–28, 54, 63–65, 70, 79, 82, 108–10, 119, 126, 129, 134, 138, 167–68, 172, 179, 193–95, 203, 208–17
Mechanized and Armor Combat Operations in Vietnam (MACOV) study: 10, 33, 84–87, 89–91, 184
Mechanized infantry: 51, 56, 64, 72, 84, 94, 96, 102, 118, 136, 140–41, 187*n*
Mekong Delta: 7, 9, 12, 21, 24, 25, 103, 106, 114, 165
Mekong River: 165, 176, 178
Memot: 172, 175–76
Midway Island: 164
Mieu Giang River: 206–10
Miles, Captain Ralph A.: 176
Military regions: Region 1, 182–83, 200–201, 203, 205, 212, 218; Region 2, 180, 182–83, 199–200, 203, 212, 215, 217; Region 3, 177, 180–83, 192*n*, 210, 217; Region 4, 180, 198, 200, 217. *See also* Corps tactical zones
Mine detectors: 81–82
Mine rollers: 82, 223
Mines, claymore: 59
Mines, enemy: 35, 47, 56, 77, 79–82, 95–96, 109, 143–44, 223
Minesweeper teams: 69, 81
Minh Thanh Road: 66, 69, 71, 107
Moc Hoa: 31–32
MONTANA RAIDER Operation: 154–55
Moore, Colonel Harold G., Jr.: 65–66
Mortar carriers: 34–36, 44, 63
Mortars: 4.2-inch, 44; 60-mm., 22, 25; 81-mm., 25, 35–36, 43–44
Mortars, enemy: 36, 62–63, 96, 97, 99
My Tho: 48, 127

Napalm: 196
National Security Study Memorandum 36: 164
New Zealand forces: 112
Nhu, Ngo Dinh: 48
Night operations: 34–37, 71–72, 94, 111, 122, 129, 141, 163

Night vision devices: 34, 144, 147
Nixon, Richard M.: 164, 174
North Vietnamese Army: 91, 114, 128, 132, 135; armor units and doctrine, 149–50, 195, 201; casualties, 136–37, 138, 180, 217; offensive of 1972, 21, 150, 164, 201–2, 205–6, 219; Soviet and Chinese equipment, 149, 193, 208–11, 213, 215–17; supply operations, 12, 95–96, 139, 153, 159–61, 166–68, 172, 174–75, 186, 190, 192, 197, 201, 203, 218. *See also* Sanctuaries, enemy; *Tet* offensive
North Vietnamese Army units: 2d Division, 133, 214; 9th Division, 176–77; 304th Division, 205; 308th Division, 205; 320th Division, 214; 324B Division, 202; B5 Front, 205; 33d Regimental Hospital, 59; 65th Regiment, 157, 159; 66th Regiment, 60; 88th Regiment, 175; 95C Regiment, 160; 202d Armored Regiment, 149, 153; 203d Tank Regiment, 214; 209th Regiment, 160; 271st Regiment, 177; 272d Regiment, 176; 812th Regiment, 145, 147
North Vietnamese government: 218

O'Brien: 146
Okinawa: 184
Otis, Lieutenant Colonel Glenn K.: 118–19, 122–23
Otis, Captain Malcolm: 122

Pacification operations: 37, 91, 114, 137*n*, 156, 161–62, 164, 180, 186
Parris, First Lieutenant William: 157
Parrot's Beak: 172
Patton, Colonel George S.: 73–74*n*, 143, 221
People's Republic of China: 149
Phan Thiet: 118
Philippine forces: 112
Phnom Penh: 175–76, 178, 197
Phu Cuong: 72
Phu Loi: 57, 60, 65, 73
Phu My District: 161
Phu Vinh: 127
Phuoc Long Province: 156
Phuoc Vinh: 156
Pich Nil Pass: 198
Pineapple Forest: 116
Plain of Reeds: 22, 29
Plei Me Special Forces Camp: 29, 66
Pleiku: 29, 107, 118, 212, 215–16
Pleiku Province: 58
Polei Kleng: 153, 214
Pon River: 196
Popular Forces, Republic of Vietnam: 140, 163, 165, 178, 186, 203*n*, 212
Porter, Colonel Daniel B.: 27
Powell, Captain Ralph: 51
Prek Klok II: 96–97
Presidential Unit Citation, U.S.: 123*n*, 163
Prey Veng: 177
Privette, Major William C.: 144
Psychological operations team: 146

Quan Loi: 155
Quang Ngai: 36, 116
Quang Tri City: 82, 145, 187, 204–5, 208, 211, 218
Quang Tri Province: 153, 161–62, 205–6
Qui Nhon: 183

Radosevich, Lieutenant Wilbert: 77
Recoilless rifles: 25, 47, 62–63, 78, 96, 97, 99, 101, 134–35
Recovery vehicles: M88, 182, 185, 204, 225; M578, 185, 225
Red Ball Express: 183
Red Devil Road: 190
Reed, Lieutenant Colonel James B.: 160
Refugees: 169, 178–79, 206
Regional Forces, Republic of Vietnam: 25, 127, 140, 163, 165, 168, 178, 186, 203*n*, 205, 212
Republic of Vietnam Air Force: 45, 167
Republic of Vietnam Army: 41, 186, 197, 199, 218; Armor Command, 17–18, 20, 22, 30–31, 33–35, 38, 63, 162–63, 193, 204–5; armor employment, 9, 21, 24, 28, 33, 37, 45, 116, 118, 127–28, 137, 179, 195–97, 217–19; armor organization, 3, 6, 17–22, 24–25, 28, 30–31, 33–35, 37–38, 136, 162–63, 203; Armor School, 17–18, 25, 29, 38, 48–49, 90, 118; command and control problems, 28, 30, 33–34, 178, 180, 192, 196–97, 211; Joint General Staff, 163, 203; morale, 137–38, 167–68, 180–81, 197; political role, 28, 34, 48–49; training, 17–18, 21, 25, 28, 33–34, 165, 197, 204, 214–15. *See also* Corps tactical zones; Military regions; Pacification operations
Republic of Vietnam Army airborne units: Airborne division, 190–96, 218; 3d Brigade, 171; 8th Battalion, 27–28
Republic of Vietnam Army armor units: 5th Armor Group, 28–29; 6th Mechanized Battle Group, 28; 1st Brigade, 162–63, 190–96, 203*n*, 208, 210–12, 218; 2d Brigade, 212; 3d Brigade, 177; 4th

Brigade, 162, 167, 197–98; 20th Tank Regiment, 203–12, 218
Republic of Vietnam Army armored cavalry units: 1st Regiment, 43, 48; 2d Regiment, 24–25, 27, 38, 43, 44, 48, 127–28, 168; 3d Regiment, 216–17; 4th Regiment, 35–37, 48, 135,210; 5th Regiment, 28, 34, 130*n*, 176, 179; 6th Regiment, 28, 31–33, 168; 7th Regiment, 156,202, 205; 8th Regiment, 33, 217; 10th Regiment,43; 11th Regiment, 136*n*, 191–93, 203*n*, 205–6, 210,218; 12th Regiment, 197–98; 14th Regiment, 212–13; 15th Regiment, 175–77; 16th Regiment, 197–98; 17th Regiment, 191–94, 208, 210, 218; 18th Regiment, 175–77, 210,218
Republic of Vietnam Army infantry units: 1st Division, 161–62, 190–91, 194–95,201, 203*n*, 205; 2d Division, 203*n*; 3d Division, 203*n*, 205,208,210; 7th Division, 21–22, 25, 27, 31–32, 165; 9th Division, 165; 21st Division, 21, 165; 22d Division, 212–14; 23d Division, 214–16; 2d Regiment, 208; 11th Regiment, 25, 27; 15th Regiment, 128; 56th Regiment, 208; 57th Regiment, 208; Hac Bao Company, 194, 197*n*
Republic of Vietnam Army Ordnance Depot, 80th: 31, 40
Republic of Vietnam Army Ranger units: 25, 29–30, 128, 163, 167, 174, 177–78, 191–92, 197–98, 216,218; 1st Group, 190; 5th Group, 211; 43d Battalion, 127
Republic of Vietnam Marine units: 29, 206,212; Marine division, 186, 190, 218; 258th Brigade, 208
Republic of Vietnam Navy: 178
Rifle, M16: 34,59
ROADRUNNER Operation: 60
Rocket launcher, 3.5-inch: 22,25
Rocket propelled grenades, enemy: 47, 97, 99, 101, 109, 116, 126, 135, 143–44, 173, 176
Rocket Ridge: 212,214
Rockets: 42, 131, 146, 159, 161, 177, 192*n*, 194, 210
Rome plow operations: 94, 147–49, 156, 160, 175–76, 199
Rowe, Chief Warrant Officer Danny B.: 216

Sagger missile, AT3: 210–11, 213
Saigon: 48, 49*n*, 91, 111, 114–15, 118–19, 123, 127–31, 156, 181, 183–84
Saigon River corridor: 91, 94, 156, 159
Sanctuaries, enemy: Cambodia, 133, 138–39, 154, 156, 159, 161, 166–68, 175–76, 178, 201, 212; Laos, 138–39, 156, 161, 166–68, 197, 201, 212
Scanlon, Captain James B.: 27
Scout cars: 3, 17–18
Seaman, Major General Jonathan O.: 57, 63, 65, 91
Searchlights: 34, 71, 94, 106, 147, 151, 187*n*, 203–4
Sensors: 80–81, 156,201
Shoemaker, Brigadier General Robert L.: 170–71
Smith, Chief Warrant Officer Edmond C.: 216
Snow, Captain Joseph: 33
Snuol: 172–74, 176–77
Song Be: 115*n*
Soviet Union: 149
Spare parts. *See* Maintenance and repair
Srok Dong: 67
Stailey, Lieutenant Colonel Raymond L.: 73, 100
STARLITE Operation: 53–54
Starry, Colonel Donn A.: 157, 160, 164, 171–73, 181, 184
Stewart, Captain Thomas: 187
Stovall, Captain John P.: 150–52
Street Without Joy: 4–5
Suoi Cat: 75, 77–78
Suoi Samat: 100–101
Suoi Sau: 104
Suoi Tre: 96, 100
Supply operation: 122, 154, 192, 218, 224–25. *See also* Maintenance and repair
Svay Rieng Province: 169–70, 175

Tam Ky: 36, 116, 133–36
Tan Canh: 212–13, 215
Tan Son Nhut: 45, 118–19, 122–23
Tank ammunition: canister, 78, 99, 102, 144, 157, 174; concrete-piercing fuze, 153; HEAT, 151-53
Tank guns: 84-mm., 113; 90-mm., 151; 152-mm., 142
Tanks: 3, 10, 72–73, 81, 94; Centurion, 113; M24, 18, 25, 44–45; M41, 29, 41, 49*n*, 112, 149, 151, 153, 191, 193, 195, 204, 212–13; M41A3, 45; M48, 51, 82, 142–43,209; M48A1, 129; M48A2C, 63; M48A3, 53, 56, 57, 63, 66, 70, 72, 129, 142–45, 203, 210–11; M80, 51; in combined arms team, 85, 97, 99–100, 102–3, 130; losses, 54, 79, 108–10, 119, 195, 210–13; in tank vs. tank combat, 149, 153, 207–9. *See also* Armor

Tanks, enemy: PT76, 150-53, 193, 208, 215-17; T54, 193, 208-9, 211, 213, 215-16; T59, 209; losses, 193-94, 208-9, 211, 215-17
Task forces: DRAGOON, 69; REMAGEN, 153-55; SPUR, 66
Tay Ninh: 111, 132, 144, 154-56
Tay Ninh Province: 95, 132, 159
Taylor, Maxwell D.: 55, 112
Tchepone: 186, 195
Tear gas clusters: 157
Terrain, effect on operations: 7, 9-10, 22, 24, 41, 64, 74-75, 85, 102-3, 105, 153-54, 169, 191, 211
Tet offensive: 21, 90, 114-16, 123*n*, 127-29, 136-38, 168, 179, 181, 201
Thach Han River: 211
Thailand forces: 112-13
Thi Tinh River: 91
Thieu, Nguyen Van: 164
Tho, Major Lam Quang: 27
Thu Duc: 49, 118
Thu Duc Military School Center: 17
Tien, Captain: 32
Toan, Major General Nguyen Van: 205-6, 208
TOAN THANG Operations: 129, 167-70, 175, 177-80, 198
Tolson, Major General John J.: 111-12
Tonle Cham: 175
TOW missile, XM26: 215-17
Tower, Lieutenant Colonel John B.: 123-24
Tra Vinh: 127
Trang Bang bridge: 119, 122
Trang Bom: 125
Troop ceilings and strength: 55-56, 58, 138, 199
Troop withdrawal operations: 164-65, 174, 176, 180, 199-200, 217
Tuyen Nhon: 32

U.S. Advanced Research Projects Agency: 38
U.S. Army Armor Agency: 86, 90*n*
U.S. Army, Pacific: 201
U.S. Army, Vietnam: 82, 89, 143
U.S. Army Armor School: 17-18, 51, 86, 90*n*
U.S. Army Aviation School: 51
U.S. Army Combat Developments Command: 84, 89
U.S. Army Concept Team in Vietnam: 31, 38, 43-44
U.S. Army Materiel Command: 89
U.S. Army Missile Command: 215
U.S. Army Mobility Equipment Research and Development Center: 82
U.S. Army Special Forces School: 19, 22
U.S. Continental Army Command: 86-87
U.S. Marine Corps: 54, 116, 139; 3d Division, 182; 3d Tank Battalion, 52-53, 164
U.S. Military Assistance Advisory Group, Vietnam: 19-21, 30
U.S. Military Assistance Command, Vietnam (MACV): 19, 30, 35, 52, 54-56, 65, 72-73, 89-90, 115, 138
U.S. Navy: 52, 103, 145-46

Van Kiep: 49
Vann, Lieutenant Colonel John P.: 27-28
Viet Cong: 5, 10, 12, 14, 16, 25, 29, 32, 45-47, 49, 60, 62-63, 97, 114, 128, 140. *See also* North Vietnamese Army; Sanctuaries, enemy
Viet Cong units: 5th Division, 132; 9th Division, 29, 66, 71, 132; 1st Regiment, 54; 271st Regiment, 67, 177; 272d Regiment, 67, 176; 514th Battalion, 103
Viet Minh: 12, 87
Vinatexco Textile Factory: 122
Vinh Long: 43, 127-28
Vinh Long Province: 127
Vinh Phuoc River: 211
Virant, Captain Leo B.: 118-19
Vo Dinh: 215
Vung Tau: 49*n*, 109, 113

War Zone C: 66, 95, 102, 132-33, 155, 159-61, 175
WASHINGTON GREEN Operation: 161
Weather, effect on operations: 7, 9-10, 12, 65, 145, 168-69, 176
West, Major General Arthur L.: 84, 91
Westmoreland, General William C.: 5*n*, 7*n*, 52, 54-58, 72, 84-85, 107, 114-15, 138, 140
Weyand, Lieutenant General Frederick C.: 58, 65, 115
Whiteis, Chief Warrant Officer Lester F.: 216
Williamson, Brigadier General Ellis W.: 65
WOLFE MOUNTAIN Operation: 161
World War II: 3, 6-7, 115, 138, 175, 220

Xuan Loc: 74

☆ U.S. GOVERNMENT PRINTING OFFICE: 2001 O- 483-088

Made in United States
Cleveland, OH
05 November 2026

47668293R00142